MESSAGE 'N A BOTTLE
The 40oz Scandal

Alfred "Coach" Powell

RENAISSANCE PRESS
CHICAGO

A secret code is hidden within the title of this book. To decode, use your intuitive and intellectual powers as you read through the next few chapters. The answer will come to you. Here's a hint: The secret code provides the solution to the 40oz scandal in the African American community. (The answer is on the final page of An African-Centered Approach to 40oz Prevention.)

MESSAGE 'N A BOTTLE
The 40oz Scandal

RENAISSANCE PRESS

© 1995 Alfred Powell
All Rights Reserved
First Edition January 1996

Renaissance Press
1507 E. 53rd St., #247
Chicago, IL 60615

ISBN: 1-888600-00-4

1st printing January 1996
2nd printing February 1996

No part of this book may be used or reproduced in any manner whatsoever without the written permission of the author, except for brief quotations embodied in critical articles, reviews, and writings on related subjects.

Edited by Donna Marie Williams
Cover design by Troy Brown
Images scanned by Firetype Typing Services

To my family-my parents Gene and Rachel Powell; my wife Shelly (a very special thank you for honoring our wedding vows, "for better or worst," during the trying duration of this project); and to Al J, Natasha, Terria, and Tiara, my loving children who sacrificed so much while I travelled the country for four years gathering information for this report. Daddy loves you very much. I pray to God that *Message 'N A Bottle* will help your generation understand the many messages of life.

To all in the African American community who were wounded or killed by the forty bombs.

To the frontline workers (on the street level) who can testify to the truth in this book.

In loving memory of
David Brian Williams
October 19, 1975 - March 19, 1994

David, thank you for helping me choose the title for this book. I know you're in heaven catching touchdown passes like you did for Dayton Paul L. Dunbar High School. Remember the song I wrote for you? You asked me to read it at your funeral service three weeks before your homegoing.

There Comes A Time

There comes a time
  when day turns to night
There comes a time
  when the night becomes the light
Before you know it, it's time to say good-bye
So I say good-bye to you with a humble sigh
So I say good-bye to you with a tear in my eye

There comes a time
  when tears of joy will flow
Those are the times
  when good friends seem to go
Before you know it, it's time to go
So I say good-bye to you with a humble sigh
So I say good-bye to you with a tear in my eye

There comes a time
  that always seems to come
Those are the times
  that take away the fun
Before you know it, the good times are gone
So I say good-bye to you with a humble sigh
So I say good-bye to you with a tear in my eye

Peace and Big Love,
Your Coach

CONTENTS

Acknowledgements *viii*

Foreword *x*

Preface *2*

Introduction *10*

1. Zima: Zomething Fishy *26*

2. Colt 45: The 40oz Harness *38*

3. Olde English 800: The British Are Coming! *60*

4. Red Dog: The Low Down Dirty Dog *80*

5. ST Ides: Satan 'N A Bottle *88*

6. Big Jug: Beware the False Prophet *112*

7. Schlitz: The Bull God Osiris Resurrected? *122*

8. Budweiser: The Final Confrontation *138*

An African-Centered Approach to Prevention *152*

Endnotes *160*

Suggested Readings *169*

ACKNOWLEDGEMENTS

After four long years of researching and writing, I must confess that this is the hardest part of the book to write. My fear is that I'll omit someone, so if I do, let me borrow from the words of Rev. Jesse Jackson and say, "Charge it to my head and not my heart."

First, all praises to God for using me as His instrument to deliver this unique message. I proudly acknowledge my connection to my African ancestors upon whose shoulders I stand without fear.

The following individuals and institutions are not responsible for the ideas written about in this book. With that disclaimer stated, let me give thanks to all of those who, for political and financial reasons (survival), couldn't afford to have their names or organizations mentioned in this book. One day we'll all be free to stand for what we believe.

This book benefitted from the love, intellectual ideals, encouragement, motivation, research, brainstorming, street interviewing, travels, hard work, financial commitment, and writing of Marlon "Shack" Shackleford. Thanks, Brother, for being there every step of the way.

A special thanks to Coach Tom Montgomery, Coach Sidney Booker, and Stan Shively for their patience and tolerance with my crazy schedules. I promise that upon my return we'll win the gold.

Thanks to my twin brother, Coach Albert Powell, for his tremendous $$$, support, and criticism. To all my other siblings, thanks for sharing the family history with me. Love ya big time.

When your head starts getting big, it takes special people to keep you in check. Thanks to Jackie Early, Darlene Powell, Victor L. Davis, Pam Thomas, Rev. El Akuchie, Elder Ernest McClellan, Celeste Taylor, Sean Walton, Randy Faison, and all of my past and present ball players.

For their constant words and acts of support, big love goes out to sisters Angela L. Cornelius, Robyn Price, Tiffany Thomas, Carla McDougle, Dr. Victoria Martin, Dr. Beverly Hall-Ogletree, and Mary Kay Earley; to brothers Roy Hollis, Ozell Earley, Leo Hayden, and Ron Allen.

A heartfelt thank you to those who refused to let us quit when the going got crazy and the bank account hit zero: Renaissance Communications and the Boss Lady Janice Williams, Darrell Imani, Mr. and Mrs. Tank Sims, Pastor Joseph T. Weston, Pastor Greg Stanton, Big Mama Frazier, Sweetheart Mary Turner Shackleford, Dr. James E. Dobbins, and Dr. Marvin Haire.

This project would have been impossible without the assistance of many persons and institutions, including Lawrence Whitman, Associate Director, Midwest Regional Center for Drug-Free Schools and Communities, and the McLendon Institute, Dayton, Ohio. Many thanks to the students of BBI/BSI (Black Brothers & Black Sisters Involvement, Dayton, Ohio) for translating street-Hip Hop lingo and completing street surveys and interviews.

The following individuals provided valuable comments and research toward the completion of *Message 'N A Bottle*: brothers Kwesi Harris, James Anyike, Khalid Fattah Griggs, Horace Fulton, A. Shakur Ahmad, Dr. Charles Smith, Ministers to Society Productions R.O.A.R., Dr. Hannibal Tirus Afrik, Al J. Powell, Rev. Mike Pfleger, Al Freeman, Harry McCloud, K.T. Harrison, Bobby Henry, Richard Green, Ron Lewis, Mandrake, Dr. Bryant Lee, Oshiyemi Adelabu, Darrell Julius, Rev. David Shirley, Ivan Rosa, Dr. Victor Lewis, Brad Franks, and Heru-Nefera. And to the sisters: Michelle Burstion, Makani Themba, Pamela Pollard, Ruth Williams, Dina Nwaokai, Hyacinth Williams, Shelly Powell, Natasha Powell, and Terria Powell.

Our excellent typists, Merrionna Pierce and Sheila Vails, translated my chicken scratch into something clean and readable. Many thanks for your hard work.

Finally, I've learned to pray carefully for God will deliver. I prayed for an outstanding editor, and He sent me the best, sister Donna Marie Williams. Thank you for being the lifeline that rescued this project. Love ya big time.

Coach Powell

FOREWORD

Keith, 16, Brooklyn, NY: "On the one hand, brothers put all their hope in the 40oz–on the other hand they put all their pain in the 40oz."

I once read that there is a big difference between rationalizing and reasoning. When you rationalize, you go from conclusion to evidence. When you reason, you go from evidence to conclusion. The evidence of my experiences has led me to conclude that Somebody is marketing 40oz of death to the Black community. Let us reason together. From the evidence that I present, see if you come to the same conclusion.

The year is 1991. I am standing alone in the middle of an empty, silent park. When I was a kid, this park used to be *the* hangout spot in the hood, but today it ain't nothing but a graveyard. Dead and dying soldiers as far as my eyes can see. Now these soldiers did not die honorably defending the hood. These soldiers were the enemy. They were dropped in my hood as an act of war. They have names like ST Ides, Colt 45, Olde English 800, and Wild Irish Rose. These dead soldiers took a lot of our own with them: potential doctors, lawyers, judges, engineers, astronauts, neurosurgeons, scientists, and fathers.

Black people often call empty liquor bottles "dead soldiers." There's a reason for that. My Uncle Rocky taught me that soldiers on leave from WWII and Vietnam would drink to drown past sorrows. They'd call the empty bottle a dead soldier. I guess the bottles were as empty as the men felt.

FOREWORD

Have you ever had a moment in your life when everything clicked and you understood with a clarity? I had one of those moments in that graveyard of a park. I picked up a 40oz bottle of Colt 45 and immediately thought about the gun that goes by the same name. Colt 45 malt liquor, Colt 45 gun. At that moment, it all came together for me. The bottle was the gun; the malt liquor, the gunpowder; the hand, the trigger. Malt liquor is *liquid* gunpowder. Every time a brother picks up a 40oz bottle of Colt 45, it's as destructive as putting a gun to his head. Coach Powell says that "Brothers go to the drive-thru,* buy the symbolic 40oz gun, then go do drive-bys."

James, 12, East St. Louis, IL: "Black youth will never be free until we free ourselves from the 40oz."

It is important to understand the role gunpowder has played in shaping the world. The Chinese invented gunpowder, which they used to light fireworks. Franz Fanon, in *Psychology of Oppression*, describes what happens when gunpowder fell into European hands:

> Europe's guns and gunpowder forced some into chattel slavery, others into sinister collaboration. The ethic of skinning others before being skinned spread like a contagion. Interracial wars became endemic, guns and gunpowder ever more indispensable. Europe gained at once more slaves, more collaborators, and a market for its new technology of death.[1]

Malcolm X, in his book *Afro-American History*, talks about how Marco Polo turned Europe on to gunpowder. The Europeans immediately began using it to kill people. Europe had

---

*Drive-thru: Many liquor stores in the Midwest have windows that you can drive up to and make a purchase without getting out of the car.

MESSAGE 'N A BOTTLE

Jay, 18, Detroit, MI: "I wake up on the 40, go to school on the 40, eat on the 40, have sex on the 40, go to sleep on the 40, but I'm not addicted to the 40."

never been able to conquer the African armies, but with gunpowder, they became almost invincible.[2]

Remember, we call malt liquor "liquid gunpowder." Think for a moment how this liquid gunpowder is affecting our community. The same way that Europe used gunpowder to subjugate a continent, alcohol companies are using malt liquor to subjugate Black people, especially those in the ghetto. This liquid gunpowder is literally blowing brothers' minds. Coach and I coined the phrase Ego Disorder to refer to the artificial, inflated ego that malt liquor creates in a brother. The Ego Disorder cycle begins with liquid gunpowder and ends with the gun.

1. Brother drinks a 40oz.
2. Brother smokes marijuana.
3. Brother gets into an argument.
4. Brother pulls a gun.
5. Brother gets shot.
6. Brother is dead–another homicide.

Rapper Chuck D of Public Enemy said on the TV show *20/20*, "There's something in the stuff that makes brothers go hostile, I mean hostile!"

One day as I was walking down Broadway in Dayton, OH, I saw seven young brothers hanging out behind Jackie's Market. They were drinking 40oz bottles of malt liquor. As I approached their circle, one of the brothers shouted out to me, "I'll give you $2.00 if you buy this 40oz for us." I figured their ages to be between six and thirteen years old. These are the young brothers people in the neighborhood and schools call antisocial and maladjusted. Bo, my nephew Dante's friend, was holding a 40oz of ST Ides and cussing like a grown man. He couldn't have been older than nine. I looked them over.

FOREWORD

"Why do you brothers drink malt liquor?" I couldn't help asking.

"Because ain't sh— else to do, so me and my boys just chillin' on the Ides," laughed Tony, 12.

Bo said he drinks because his moms and her boyfriend drink. "They up there drinkin' now," he said.

Cal, 10, drinks because it makes him feel good–like he can do anything. He said, "My moms gave me my first drink when I was six years old."

"My mama said ain't nothin' wrong with drinking," said Bo. "It can't do nothin' to you. Can it?"

Bo's cousin JD, 13, sniffed and tugged on his pants. "You OK as long as you stick by me." Tony added that his sister drinks whenever times get hard. Drinking helps her to think and forget her problems.

I explained to the brothers that alcohol and cigarette billboards, liquor stores, pawn shops, and street shops* that sell guns and drugs are strategically placed in our neighborhoods. I offered to give them a guided tour. They hopped in my van, and as we rode, I talked to them about alcohol being liquid dope and the #1 legal drug on the market. "Alcohol is a multi-billion dollar business," I said. I asked them if they remembered Malcolm X's line in Spike Lee's movie: "Every time you break the seal on a liquor bottle, you're breaking the government's seal."

I told them I could prove Malcolm's words. I had them look on the side of the ST Ides bottle. It read: "Government warning: (1) According to the Surgeon General, women should not drink alcohol beverages during pregnancy because of the risk of birth defects. (2) Consumption of alcoholic

Robert, 15, Davenport, IA:
"Us brothers from the suburbs drink just as many 40s and smoke as much weed as brothers from the inner city, if not more."

---

*Street shops: Places in the hood where you can buy illegal items.

Ronnell, 18, Dayton, OH: "Even before we lose our brothers to prisons and cemeteries, it seems we lose them to the 40oz first."

beverages impairs your ability to drive a car or operate machinery, and may cause health problems." Keith, 11, asked, "If the Surgeon General says it's bad for your health, why would they sell it to you?"

The government has the audacity to say that drinking alcohol *may* cause health problems! That warning label should read, "(1) According to the elders in the Black community, alcohol *will* cause sexual abuse, child abuse, 50% of all homicides, teen pregnancy, child molestation, date rape, suicide, incest, and rape. (2) Consumption of alcohol beverages *will* cause alcoholic hepatitis, cancer of the esophagus, cancer of the larynx, cirrhosis, fatty liver, stomach ulcers, lung cancer, infant mortality, and fetal alcohol syndome."

"We didn't know that," said Keith. "Are you sure alcohol does all of that?"

Every time you (the consumer) buy malt liquor, the investor makes money. Some neighborhood stores will actually sell this liquid dope illegally to minors just to make more money. There's a conspiracy to produce child consumers of alcohol to ensure transgenerational drinking. As we rode down the streets in the hood, we found more than 15 billboards promoting 40oz malt liquors within a six mile radius. Then we went on to Oakwood, a suburb of Dayton. We couldn't find any billboards promoting malt liquor. While in Oakwood we also went to three different stores looking for ST Ides. Couldn't find narry an ounce.

JD and Shawn asked, "Man, why they got 40oz in the hood, but not here?"

"NOW YOU'RE THINKING!" I said.

We went back to the hood, and I took my awakening crew to some drive-thrus. We looked at the many ad posters that exploited the sexuality of Black women. Even the names of the malt liquors were seductive. Power Master. Mustang.

FOREWORD

The subliminal message is, "If you drink this, sex will follow." Wilson Bryan Key coined the phrase "subliminal seduction." When you drink alcohol, it's not you talking but the alcohol through your subconscious mind. Alcohol talks to you, with you, and through you. The 40oz can change Dr. Jekyll into Mr. Hyde.

"What animal is on the label of Olde English 800?" I asked the young brothers.

"A tiger," they said.

"And a tiger is what–powerful or powerless?"

"Powerful."

"There's also a malt liquor called Midnight Dragon," said JD.

"And a dragon is what?" I asked.

"Powerful!"

"Plus it breathes fire," said Keith.

"Sounds like some firewater to me," I said.

Lisa, 20, a sophomore at Central State University: "The 40oz is treated like a god by men and women on campus."

• • •

Alcohol is like liquid crack. It has brothers cracking up and breaking up. My Uncle Rocky used to say, "Gin will make you sin" and "Wine will make you commit a crime." He called alcohol "evil spirits." He said, "They sell it to you, then when you get drunk, they lock you up. They make more, then they let you go and sell it to you again." He laughed and said, "Alcohol taxes is big business." Even though my uncle was aware of the causes and effects of alcohol, he died a slave to the drink. His last words to me were, "Watch out for the 40oz."

That's just what Coach Powell has been doing–watching out. *Message 'N A Bottle* is his dossier on the marketing cons of alcohol companies that sell malt liquor. He runs down the cultural and religious themes that are put on malt liquor

MESSAGE 'N A BOTTLE

Cynthia, 14, gang member, Los Angeles, CA: "The 40oz is part of the initiation for our set. We love the sh– too."

bottles and ads. He peels the labels off the bottles and holds them up to the light of day. It's hard to keep a con going in the sunlight. His information will make the blind see and arm the seers with strategies to expose and fight the 40oz scandal.

At Black Brothers & Sisters, Inc. (BBI/BSI), we don't look at alcohol consumption as legal vs. illegal. Obviously if you're under 21, it's illegal. As BBI/BSI advisor Sean Walton once said, "We have to look at it as legal vs. *immoral*." It is immoral when a 40oz party leads to violence and a brother or a sister pays with his or her life. It's immoral that alcohol is the number one drug in our community and that it creates more business than funeral homes, hospitals, treatment centers, liquor stores, and drive-thrus can handle. It's immoral when brothers keep going even when the sisters say no. It's immoral when sisters wake up so hung over they haven't a clue about what happened the night before. It's immoral that brothers get behind the wheel of a car not knowing how to add: 40oz malt liquor + 20 mph in a school zone = 60 mph of danger to our children. It's immoral that Black leaders and Black media sacrifice the sanity of our community for a few pennies from alcohol companies. It's immoral that students spend $5.5 billion on alcohol annually, more than they spend on nonalcoholic drinks and books combined.[3]

Can you imagine today's alcoholic college students becoming tomorrow's doctors, lawyers, nurses, teachers, police officers, and preachers? Instead of getting down with the books, too many of our young people are doing beer runs. Alcohol has become the cult inside Black culture. In order to fight Black-on-Black violence, we must fight alcohol-on-Black violence. The 40oz is Public Enemy Number One. Malt liquor has initiated its own rites of passage program. The 40oz is for childhood; the 45oz, boyhood; and the 64oz,

FOREWORD

manhood. As Coach Powell says, "The door to manhood and womanhood should not be the cap on a 40oz bottle!"

The Black community must become a detox center. Our people must take the initiative in fighting alcoholism in our community.

Why should you read *Message 'N A Bottle?* This book raises awareness, the first step in busting up the 40oz plague in our community. Our deepest hope is that this book will open up and contribute to the dialogue in schools, churches, prisons, colleges, and on the street. When so many young people in their prime are starting to develop a relationship with the 40oz, we need to start talking, planning, and acting *now*.

America promised us 40 acres and a mule. Instead, we took to the 40oz like fools. There are too many dead soldiers in our community. We need live soldiers to fight for Maat- truth, justice, and righteousness. Let the habituation* begin.

Marlon "Shack" Shackleford
Black Brothers & Sisters, Inc.
June 1995

*As **Keith**, 15, Miami, FL:*
"Me and my hoe drinks four 40ozs a day. I rather drink 40oz than water!"

---

*Habituation: incorporating new habits.

MESSAGE 'N A BOTTLE
The 40oz Scandal

PREFACE

It has been estimated that the alcohol industry reinvests less than $.25 for every $25,000 they make off of the African American community. Obviously our Black lives ain't worth a quarter to them, dead or alive! The only cents the African American community should spend on malt liquor (alcohol) is our "common sense."[4]
—Paul Kelly, Director of Alcohol and Drug Prevention, Bobby Wright Mental Health Center, and co-founder of the City-Wide Coalition Against Tobacco and Alcohol Billboards (Chicago, IL)

I invite you, better yet, I dare you, to come and take a walk with me through the hood–any inner city community in Cleveland, Chicago, Washington, DC, Atlanta, Milwaukee, Boston, Philadelphia, Compton, Little Rock, Harlem, Oakland, Detroit, Los Angeles, Newark, New Orleans, Buffalo, East St. Louis, Portland, Dallas, Houston, Winston Salem, Denver, Baltimore, Dayton, or Kansas City. Believe me when I say, the scene is the same wherever you go!

It won't be long before you recognize that *they* are everywhere. Of course I'm referring to the 40 ounce (or as we say in this book, 40oz) bottles.

Look to your left, now to your right. A 40 over here, a 40 over there. You get the feeling that everybody's doing that "forty up" thang (drinking malt liquor). Brothers and sisters, young and old are holding impromptu 40oz conventions on street corners, outside storefronts, in the alleys, and

PREFACE

while riding four deep in their cars.

Empty bottles are everywhere–near school playgrounds, housing projects, parks, under abandoned cars, and in church parking lots. You wonder why these oversized 40oz bottles, entombed in their brown paper bags, never make it to the trash can. Lying empty on the ground like that, the empty bottles cleverly advertise the deadly message: we saw, we bought, we drank, and it was good.

Look around the corner and you'll see a group of young brothers and sisters with 40s in their hands. But isn't today a school day? you ask. Yes, it is. Some of our young people are so addicted to malt liquor, they would rather drink than learn. For example, check how one of the brothers stuffs his cigar with a little weed (marijuana), then dips it into his 40oz bottle. They call that "cookies and milk" or "B-40" (Bomb 40).

Who's selling this liquid dope to our young people? Isn't that illegal? you ask. Yes, it is. Walk down the block with me to the corner store owned by foreigners (pick any country) who could give a care about our community. They are selling malt liquor to obviously underaged brothers and sisters. No questions asked.

See that sister over there? She just dropped her children off at school. It's 8:15am and already she's doing "80 mph." That's slang for smoking weed (from a 40 ounce bag), then chasing it with a big 40oz guzzle of malt liquor.

See that crowd over there standing in a circle? For them, drinking the 40 is a rite of passage. The game they are playing is called Spin & Swallow, a playoff of Spin the Bottle—but make no mistake, this is no childhood game. A group of youth stand or sit in a circle and spin a 40, 45, or 64oz bottle of malt liquor. Whomever the bottle stops and points to has to take a big swallow. Normally, the group will finish off four to ten bottles before the game is over.

There's a house party almost any night of the week. We're crashing the one on 45th Street. As you mix and mingle, check out the lingo of Black males as they drink in groups. The psychology is mind blowing. Brother Shack and I have listened to the brothers, young and old, across the country, and the language is always the same.

"Go 'head, bro, *you kill it.*" Kill what? you ask. On the surface, the phrase means to take the last swallow from the bottle. But let's take it deeper. Brothers & Sisters, alcohol is called *spirits*. Obviously, malt liquor is an evil spirit because of what it does to the body and mind.

The language strongly suggests that on some level of consciousness, the brothers understand that something is alive in the bottle, something evil. It's as if they are hoping that someone will take on the duty of killing this evil spirit before it kills them! No brother or sister has won this one-on-one battle to date, and the alcohol industry continues to introduce bigger (40oz, 45oz, 64oz, etc.) spirits to the Black community.

This party's really going on. Dancing, bid whist. Let's watch as the brothers and sisters play Tap the Bottle. The point of this game is to drink three to five 40oz bottles of malt liquor each.

The next morning the sun rises over the alley, a graveyard of empty 40s. Think of the great Black men in our history such as Hannibal, Rev. Nat Turner, Gabriel Prosser, and Rev. Denmark Vessey, and how their offspring and blood-

PREFACE

lines have been poisoned.

Does everyone in the hood spend their time drinking and getting high? you ask. Absolutely not. This love (addiction) for 40s and a blunt (marijuana) belongs to a "hard-core" subculture. We are all connected, so what happens to some affects us all. Our fear is that this fad is becoming more acceptable within the mainstream of our culture.

To make a bad problem really bad, Spin & Swallow and Tap the Bottle are now being played on Black college campuses across the country. This should not be a major surprise given the fact that hip hop and so-called gangsta rap music and videos glorify the 40oz. A hip hop commercial for Sprite even suggests that the soda should come in a new 40oz container.

A magazine named *Forty Ounces & A Blunt* intensifies the glorification. The owners defend their name with the following statement:

> *40 Ounces & A Blunt* is a reflection of the street; a mirror image. Whether right or wrong, the symbolism of beer and illegal herbs as a title is a reality of life in our generation. We at the *40* are committed to presenting and preserving that reality at all costs.[5]

I hope that *Message 'N A Bottle* will make a change in this generation's sad reality of life. We believe that the promotion of the 40oz (and a blunt) is a genocidal plot to kill the Nation of Hip Hop. Are we too high to see?

Quick, look to your left. See those young brothers and sisters standing in front of that van? They're being asked the following questions:

- What type of music do you listen to? Who's your favorite musical artist?

Message 'N A Bottle

- What's your favorite malt liquor?
- Do you smoke cigarettes, cigars, or marijuana?
- What's your favorite size of malt liquor?
- Would you buy malt liquor in a bigger size?
- Do you drink alone or with friends?
- Would you like to test sample our newest malt liquor?

The young people walk away with free rap and hip hop tapes, posters, and tee-shirts. They think they've gotten over, but what they fail to realize is that they've just played informer. They've given away valuable information about their lives for mere trinkets. The corporate "brothers" and "sisters" come down here to collect vital cultural information about our language (slang), street psychology, the Black male ego, musical tastes, employment status, fashion trends, health status (mental and physical), level of violence in the community, sexual practices, diet, sleeping habits, etc. This information is used by the alcohol industry to scientifically target groups such as Blacks, Hispanics, native peoples, gays, and women.

Exploitation is the name of the game.

Have you seen Spike Lee's movie *Drop Squad*? If not, you need to rent it immediately. The movie is about a Black advertising executive who designs ads that exploit Black people. One ad was for "Mumblin' Jack Malt Liquor," and had the tagline, "MJ gets ya cozy!" Mumblin' Jack Malt Liquor came in a *125*oz bottle. Brothers & Sisters, I believe this is prophecy. It won't be long before the 125oz bottle is introduced to the Black community.

Let's go to the park. Look at all those empty bottles by the swings. Could it be a coincidence that the sizes of those bottles bear a suspicious resemblance to certain makes of guns that are popular in America? Consider the following breakdown:

PREFACE

>   12oz = 12 gauge shotgun
>   16oz = M16 machine gun
>   22oz = 22 caliber
>   32oz = Saturday night special
>   40oz = 40 gauge shotgun
>   45oz = 45 magnum
>   64oz = double barrel shotgun (two 32ozs combined)
>   125oz = atomic bomb

Every time you drink from these bottles you're symbolically blowing your brains out.

Be careful as you walk. I wouldn't want you to step on any of those dead spirits (alcohol). Now read the names of those spirits: the slave trader's **Olde English 800**, the false apostle **ST Ides**, the powerful **Colt 45**, the lewd **Red Dog**, the blasphemous **Zima**, **Big Jug**, the false prophet, **Budweiser**, the messenger, and **Schlitz**, the bull god.

See those fine women on billboards, liquor store posters, and point-of-purchase advertising? The little brothers are sure checking them out. You can't walk through the hood without seeing women selling alcohol, beer, and malt liquor. Sometimes you'll find these and other ads close to schools–right where children can get a good look at them.

Read those million dollar taglines: "The Bull is taking charge." "It's a Bud thing." "Zomething different." The one for G. Heileman Ex-Imported Beer says, "The joy of Ex." This slogan plays off the title of a very popular book, *The Joy of Sex*, thus linking sex to alcohol consumption. Another reads, "Men think about Ex 100 times a day." Which men think about Ex (sex) 100 times a day? Given that this billboard ad is mostly seen in Black neighborhoods, we can conclude that G. Heileman is insinuating that *Black* men think about sex 100 times a day. More stereotypes, more myths, more lies.

These ads frequently use sexual themes, images, and subliminals to get inner city youth to buy buy buy. What the ads don't say, however, is that with alcohol being a depressant, the sexual response, especially in males, is numbed. We're talking impotence. Those men who do think about Ex 100 times a day perhaps are unaware of the numerous studies that prove alcohol's negative effect on sexual performance.

The ability to make good decisions is compromised under the influence. **Sex + alcohol use = unwanted pregnancies and STDs**. In fact, alcohol lowers inhibitions so that violence becomes easy to commit. **Violence + sex (even impotence) = rape, gang rape, date rape**.

We're coming to the end of our tour. See that house over there? It's a popular after hours bootleg joint. From 2:00 to 5:00am you can get all the alcohol you want–if you're of age. Bootleggers in the hood, unlike the corner liquor store owner, won't sell alcohol to teens. It's a rule of the community.

Speaking of liquor stores, have you ever seen so many in your life?

> When last year's riot erupted, South Central had a staggering 728 licensed liquor outlets— 13 per square mile. South Central had more stores selling hard liquor than 13 entire states can claim and a liquor outlet at virtually every major intersection.[6]

The alcohol industry is relentless in their targeting of Blacks. While activists are protesting against the 40oz, the industry is busy introducing bigger bottles and more potent brews.

PREFACE

When you go back to your homes, offices, and classrooms, tell your colleagues, family, and friends that you took a walk through a war zone today. Tell them that you saw the victims and frontline soldiers up close. And give them this message: We need reinforcements and ammunition because we're fighting back!

Coach Powell
September 1995

INTRODUCTION

"Beer advertising and marketing materials should not employ any symbol, language, music, gesture, or cartoon character that is intended to appeal primarily to persons below the legal purchase age. Advertising or marketing material has a "primary appeal" to persons under the legal purchase age if it has special attractiveness to such persons above and beyond the general attractiveness it has for persons above the legal purchase age, including young adults above the legal purchase age."
–Beer Institute Advertising and Marketing Code

Message 'N A Bottle: The 40oz Scandal is my small contribution to the overall Master Plan to counter the conspiracy to purposefully intoxicate and destroy Black families and communities. Most of the research, prevention, and treatment around alcoholism in the Black community are being done on educational, institutional, and governmental levels. This book takes a look at the problem from a grassroots level. Our research, analysis, and writing are all from the grassroots.

I make no claims to objectivity. How can you be cold and calculating when your people are being destroyed? I do understand that many cultures in America are being targeted by the alcohol industry–women, gays, and youth of all ethnic persuasions, and I disagree with the deliberate targeting of any group for destruction. However, I chose to dedicate my first written report to the group I consider to be most at risk: Black youth.*

The most popular drugs in the youth community are alcohol and marijuana (weed, blunt). These two drugs are

---

Message 'N A Bottle, Volume 2, will look at the alcohol industry's targeting of youth from a multicultural perspective.

INTRODUCTION

called gateway drugs because they often lead to addiction and use of harder drugs such as cocaine, crack-cocaine, LSD, and heroin.* In general, rates of alcohol and marijuana use among Black youth are less than other groups, however, some surveys (PRIDE 1993-94, 1993 National Household Survey on Drug Abuse) reveal that those numbers are drastically increasing among both males and females.

As for rates of malt liquor use specifically, my guess is that the marketing research departments of the alcohol companies are keeping those numbers close. Some companies say that they don't count youth numbers, but we don't believe it. As you'll discover in the following chapters, the campaigns are too youth oriented for them *not* to track the effects. From a purely business perspective, not counting youth numbers would not make good business sense. The more alcohol companies know about our young people, the better they'll be at selling them products. "Know thy consumer" is the commandment of business, and judging from what Shack and I hear on the streets, these companies know our children too well. We know from our own work with Black youth and from our interviews with community activists, educators, parents, ministers, and others that malt liquor use among Black youth is on the rise. Malt liquor ads and promotions, combined with dismal social conditions, are affecting the minds of our children. They're taking to the 40oz like babies to their bottles.

Look, "each year, more than 1.1 billion cans of beer are consumed by junior and senior high schools, according to the

*The latest gateway drugs on the scene are inhalants, which are basically alcohol-based aerosol products sold over the counter (e.g., hairspray). You don't need an ID to buy them, and compared to alcohol and other drugs, they are a relatively cheap (and dangerous) high.

Question: What does crack really mean to Black people in America?

C = the conscious conspiracy committed upon communities of color via the use of cheap cocaine made to cause confusion, chaos, and community-cide

R = a radical form of racism, to reduce, remove, and redefine a race of people (guess what race) to repress the ideals of reparation and revolution

A = the attempted assassination of every African American and all those who associate with them via the use of AIDS and academic suicide (is this the American way?)

C = cold, calculated chemical, biocidal, psychological, physiological warfare committed against communities of color

K = the killing of the soul, the dream, the dreamer, the belief the believer, the kings and queens of Kamit, which you know as Egypt on the continent of Africa whose true name is Alkebulan.

Brothers & Sisters, that's the real breakdown of CRACK. You'd better ask somebody, then tell somebody.

"It is estimated that Blacks spend $11 to 12 billion per year on alcohol alone."
–Lawford Goddard, Director, Education and Training Institute for the Advanced Study of Black Family Life

US Department of Health and Human Services, translating into $200 million in revenue for the beer industry."[7] The malt liquor segment of the beer industry survives because of the enthusiastic patronage of African American youth. During the 1980s, malt liquor ad campaigns started to take on a distinctively hip hop flavor. Run DMC for Olde English 800. King Tee for ST Ides. In 1991, the California-based Marin Institute released a media action alert on the deliberate targeting of malt liquor to African American youth. At the time, beer sales had been declining, but malt liquor sales had increased 11%. "According to the Institute on Black Chemical Abuse, African-Americans consume one-third of all malt liquor although they constitute only 14% of the population. The G. Heileman Brewing Company estimates that African-Americans consume 75% of Heileman's Colt 45."[8] As we at the grassroots level know only too well, and as we show throughout this book, Black youth, especially in urban areas, form a sizeable percentage of malt liquor drinkers.

> People who work daily with Cincinnati area teenagers in treatment for drug and alcohol abuse report most of their clients have replaced plain old beer. . .with the stronger and cheaper malt liquors . . .The result: kids are getting sicker more quickly and addicted sooner.[9]

This book was written for those of us working on the street level with Black youth, as well as for ivory tower researchers and educators, government funded institutes and think tanks, law enforcement, juvenile justice, the medical profession, community-based agencies and organizations, parents, families, educators, policy makers, businesses, media, and faith communities. Most of all, *Message 'N A Bottle* was written for the young people who are struggling with how to

INTRODUCTION

"just say no" in a "say yes to alcohol" world. Mixed messages from media, family, peers, and heroes have turned this generation into a highly sophisticated, yet world-weary and confused group.

My contribution to the struggle involves exposing one of the best kept secrets in the alcohol industry–the many underhanded, slick ways a particularly obnoxious upstrength brew is being sold to Black youth. Once the sexy, cultural messages take hold of their minds, they just can't seem to let the bottles go. It's as if malt liquor was specifically designed for us, for our biochemical psycho-social makeup. It's surely being aimed at us.

What's the big deal? It's only beer!

> My scientists have come up with a drug that can be smuggled into their brews to effect slow poisoning results and fertility destruction.
> –former South African President P.W. Botha (August 18, 1985 edition of *Sunday Times*, a South African newspaper)

Malt liquor *is* beer, however, the addition of sugar during the brewing process increases the potency. Ice beer, which was dropped on us in 1993, gets its high potency from freezing batches of brew then removing the ice crystals. Regular beer averages about 4% alcohol by volume, and ice beer can go as high as 4.6%. Without a doubt, however, malt liquor wins the prize, with some brands as high as 8%. That's twice the alcohol content of regular beer.

Abuse of alcoholic concoctions that are loaded with sugar (malt liquor and coolers) can lead to alcoholism which, in addition to psychological misery, can lead to hypoglycemia. In addition to sugar, malt liquor contains low levels of grain

"The Fanon Center in Los Angeles estimates that 16% of the African American population is alcoholic. An estimated $16 to 21 billion is spent on treatment alone."
–Lawford Goddard, Director, Education and Training Institute for the Advanced Study of Black Family Life

alcohol. Grain alcohol is also used to make crack cocaine. No wonder teens call malt liquor "liquid crack."

Malt liquor is bad enough, but when ". . .the 40-ounce size was introduced in the mid-1980s, national malt liquor consumption [rose] from 73.6 million of the 2.5-gallon cases in 1989 to 82.9 million of the cases last year, the New York Times reported."[10]

The 40oz bottles are sold and promoted as one serving. One 40oz bottle is equal to three 12-ounce cans of beer and five shots of whiskey. The oversized bottles definitely encourage binge drinking. What brother do you know shares a 40? Very few.

As the popularity of the 40oz grows, so will the size of the bottle. Don't be surprised when the alcohol industry drops the 80 and 125oz bombs on the Black community. I can hear some sellout fake gangsta rapper singing now:

If you wanna get the ladies
pop the cap on a 80
If you wanna pop that behind
buy a 125.

The billboard slogan would probably read, "Bigger is betta!" You'd better ask somebody!

Simply put, the oversized servings of malt liquor promoted to our young are designed to kill them faster, cripple their spirits, break down their bodies, destroy their abilities, reduce their greatness, and encourage disobedience to the Creator. Some people argue that it makes no sense for the alcohol industry to kill off their profit line, and I reply, "When has genocide ever made sense?"

For 16 years, I have had the fortunate opportunity and pleasure to coach some of America's best student athletes.

INTRODUCTION

There is nothing finer than a young Black man who is fit academically, physically, psychologically, and spiritually. At his best, he can display great feats in track and field and all team sports. I have watched young Black men under tremendous competitive pressure beat the odds.

As strong as these young men are, however, they can fall like Goliath when shot down by the 40oz. I first became aware of the 40oz bombs that were being dropped on my community during the summer of 1988. My students, loved ones, friends, and neighbors were the targets. Like other coaches in the league, I began to lose student athletes before they had a chance to get started in life.

In 1992, my colleague Carla McDougle coaxed me into talking about drugs and violence to a group of eighth graders. What I learned from these young brothers was that they didn't have any real information about malt liquor to make an intelligent decision about whether or not to drink.

"A couple of 40s won't hurt nobody." "It's just beer, man." "Drinking 40s won't make you an alcoholic."

I became convinced that just saying no just didn't cut it. These young people had been getting their information from each other, the older brothers, alcohol ads, and the media. Nobody was telling them the truth.

My mentor, brother Ernest McClellan, told me that if I was going to take on this war, I would have to fight fire with water. He said that water was the symbol for truth, peace, and life, and that anytime you apply those virtues to a problem, truth and justice can't help but spring forward. He said if the young brothers love the 40 so much, let them know that the 40 doesn't love them. Don't tell them, show them, he said. Talk on their level. Teach them how to say "hell no" to the 40 and still save face. He told me not to sugarcoat the information because sugar is toxic.

Culture Rape

Clearly, the psychoculture of urban Black America has become a raping ground for the alcohol industry and those who seek profits, not promise; war not peace; Satan, not God; gangs, not healthy families; separation, not togetherness; ghettoes, not communities; and projects, not homes.

Many years of coaching football has taught me how to recognize complex schemes. It's clear to me that the alcohol industry has unleashed a very complex offensive against minority communities. With one hand they give us money for scholarships, community-based programs, sporting events, and concerts. Brothers & Sisters, use a little common sense here. Do you think the alcohol industry would continue to support and fund Black events, culture, and higher learning if the Black community stopped drinking alcohol? OF COURSE NOT! Clearly, that makes our relationship with them a pimp-whore affair.

When will we stop bending over and stand up as a people and do for ourselves? Throw out this welfare relationship we have with the alcohol industry! They say we all have a price, but is a few bucks worth the health and sanity of our children?

I said the alcohol industry's scheme is complex. With the other hand, they're making the death exchange-your money for their dope. They act like they care, but they really want us addicted to their products. The alcohol industry only markets to two groups of people: alcoholics and potential alcoholics. No one escapes.

Have you ever noticed that malt liquor, coolers, and upstrength wines are cheap? Is it a coincidence that malt liquor is targeted to poor communities? Seems like these companies are dollar dependent on our addiction to alcohol. For example, Black consumption of ST Ides literally saved

INTRODUCTION

the parent company of G. Heileman Brewing Company from bankruptcy.[11]

"Philanthropy" is a deep con, but what really gets the attention of young and old alike are the advertising campaigns. They make us laugh. They make us snap our fingers to beats that we hear everyday on our favorite Black radio stations–R&B, reggae, funk, and rap. Our myths, folklore, art, literature, history, symbols, rituals, language, music, customs, fashion, and religion are all used to suggest the idea that reckless drinking is a Black cultural thing.

Have you noticed that in urban communities, malt liquor ads are everywhere? Obviously, there is a conspiracy to keep us from going into the 21st century healthy, sober, and sane. "Studies in Baltimore, Detroit, New Orleans, Washington, DC, St. Louis, San Francisco, Philadelphia and Atlanta found that from *50% to 75%* of the billboards in low income and/or Black neighborhoods were for alcohol and tobacco, compared to only 20-36% in White neighborhoods."[12]

By now, the evil twin has found its way into your child's body. I'm talking about the alcohol product look-alikes–fruit juice drinks, ice teas, and sodas–that use similar logos, bottle shapes, and salesmen superstars. Traditionally, alcohol, tobacco, and even marijuana have been considered gateways to harder drugs, but with the sinister packaging of these nonalcoholic drinks, we must rethink what a gateway substance really is. I believe we should expand the definition to include anything, including candy cigarettes, that might lead to substance use. Brother Marlon Shackleford states,

> Corporate America and Madison Avenue continue to play dirty. They are hitting us below the belt. They are using the scheme of Psychomediaperpetrator Disorder, which can be defined as

The September 4, 1995 issue of Advertising Age *reported that a new ice tea was launched north of the border in Toronto, Canada, and it's being marketed to the young adults of Generation X. "Unlike the potent strain named for Long Island, Atomic Ice Tea's alcohol comes from the fermentation of hops and yeast rather than spirits, Molson [Breweries] said. But Atomic packs a punch, as print ads buy BBDO, Toronto, a test with the no-nonsense tagline: 'Atomic Ice Tea. 4.5% alcohol.' " Ice tea today, alcohol tomorrow.*

monkey see, monkey do. They are taking advantage of our children's impressionability and vulnerability. I bet if you dig long enough you will find a common thread that will connect the owners of some of these alcohol products to the owners of the fruit and ice tea drinks, and vice versa.

Brothers & Sisters, we are being played like fools. Everfresh Juice Co. sells its Everfresh Tea in bottles identical to bottles containing Lawrence Grapefruit & Gin. Same with MD 20/20 wine and Night Train Express. The hottest song out during the summer of '94 was Snoop Doggy Dogg's *Gin & Juice*. Snoop Doggy Dogg is a pitchman for ST Ides malt liquor by night and Crooked I fruit punch by day. Both products share the same crooked letter "I" logo and a similar color scheme. When young brothers in the hood buy a bottle of Everfresh Ice Tea they peel the label off the bottle to make it appear as if they're drinking from a bottle of gin.

Those on the frontlines know how hard it is to compete with the alcohol industry. We both want to influence the same young, intelligent Black minds. The alcohol industry strongly denies that they are targeting urban youth, but their advertising campaigns suggest otherwise. Their ads appeal to the sex ego of males and females, and they use our language to do it.

> "Get your jimmy thicker."
> "It's a Bud thing."
> "You betta recognize."
> "Down with M.G.D."

Malt liquor advertisers use many of our musical geniuses and entertainers to pitch their poison. The same brothers and sisters who rap about **revolution** support the alcohol industry's apparent belief that Blacks are **niggas** who just love to drink

INTRODUCTION

and get high, and since the world knows that **niggas** are worthless and lazy, that makes it OK to sell them killer products. Hello.

I know, it's not personal. It's just business—the business of killing your hopes and dreams. And if you die in the process, so be it.

Many of the malt liquor ad campaigns have an outlaw aura about them. This is done to appeal to the so-called wannabe gangsta ruff neck. Perhaps the malt liquor industry thinks it's their social and moral duty to kill gangstas. Thus, when their product kills you or your children, no one complains because it was "justifiable homicide."

Reading Between the Lines

America's favorite Black maid has been given a makeover on a pancake box, while on another, an old Black man still fetches cereal for his master. A white man in blackface sells American toothpaste overseas. In a past issue of a leading company's employee magazine, Africans were depicted as monkeys, while other countries were given human representation. What are these symbols from America's slavery past really telling us today? Has anything changed?

It's important that we define what we mean by symbol. I find Jung's definition helpful.

> A symbol is a term, a name, or even a picture that may be familiar in daily life, yet that possesses specific connotations in addition to its conventional and obvious meaning. It implies something vague, unknown, or hidden from us.[13]

We may *see* an image or read a word or a phrase, but we *think* and *feel* on many, many levels. That's what is meant by the saying, "One picture is worth a thousand words." Books written by Wilson Bryan Key and Frances Cress Welsing explore how symbols are used in media and popular culture. I encourage you to read all of their books to gain an excellent understanding of how symbols are used to manipulate minds and spending behaviors in American society.

Straightforward advertising would be hard enough to combat, but at least we'd know what we were fighting. What makes the malt liquor ads so hard to fight (and resist) are the use of subliminally embedded symbols taken from the cultures and histories of various societies, including the African American. Consciously we may be tapping our feet to the beat, but subconsciously, our minds are both disgusted by and attracted to the use of life, death, religious, and sexual symbols.

When I talk to audiences, the hardest thing to overcome is the disbelief of the people. It's hard for people to admit that they've been conned in such an underhanded, evil way. American consumers have been so conditioned to accept what's on the surface of advertising that to suggest that our most sacred symbols are lurking within the pictures and copy is ludicrous to many people. You'd think folk had never taken a college literature or art class. For centuries, Western art has been characterized by its symbology. Nothing is what it appears to be. Everything stands for something else. With the advertising industry fueled by well paid, highly skilled writers, artists, psychologists, researchers, technological wizards, and cinematographers, many of whom have been trained in traditional liberal arts/humanities courses of study, is it so ridiculous to assume that they have carried on the tradition of Brahms, DaVinci, and Poe?

Message 'N A Bottle unscrambles the mixed messages found in the ad campaigns of Zima Clear Malt, Colt 45, Olde

INTRODUCTION

English 800, Red Dog, ST Ides, Big Jug, Schlitz, and Budweiser. We look at the symbols used by malt liquor advertisers and decode some possible meanings on cultural, spiritual, physical, and psychological levels. African tradition says that your name must have meaning, so we look at the names of the products and the meanings behind them. This book helps the reader to solve many of the riddles cleverly disguised as innocent slogans. We examine each slogan and symbol from both a literal and figurative perspective. We pick apart those symbols, images, language patterns, slogans, and musical beats that are pumped past our subconscious mind (eyes and ears) at lightning speed so as to bombard our collective psyche.

The hidden messages found in many of the ad campaigns and labels are so mind-blowing that I honestly found myself frightened and at one point considered stopping the process of writing this report. Slavery, rape, sex, crucifixion, castration, and satanic themes run throughout most of these ads. But my faith in God and my support team told me we had come too far to turn back now.

To help us in our study, we used a simple process we call *you*cation, which means that you put yourself in the middle of the information, then examine it using your own life experiences, culture, race, gender, history, environment, and customs. This is why *Message 'N A Bottle* comes from an African-centered, Christian perspective. It's who I am, it's how I try to live my life. I also feel more than justified coming at it from this angle given the fact that most of the malt liquor ads are being targeted to the Black community. The African-centered framework of our study allows us to openly discuss the historical legacy that racism, egotism, sexism, and poorism play in the addiction habits and patterns of African people in this country. Brothers & Sisters, it's important to understand that most, if not all, of the music, symbols, signs, and slogans used to sell malt liquor to African Americans have been stolen from the Black experience. Black spirituality, Black culture, Black history, and Black sexuality are being exploited. And because Black culture is revered and imitated by other races and cultures, the malt liquor ads reach a wide variety of people.

Many people ask me how I came to understand and recognize the hidden messages. First of all, I don't think the messages are hidden. I see them as scrambled. Secondly, 10% is study, another 10% is luck, and the rest is divine inspiration. I call the process *decoding*.

To begin the decoding process, I start with *you*cation, that is, I review all the symbols, language patterns, music, etc., from my personal perspective, which is African-centered. Secondly, I apply the rules from several board games and TV game shows such as *Scrabble, Jeopardy, Match Game, Classic Concentration*, and *Name That Tune*.

Suppose I told you that the word *sex* and an abstract swastika can be found inside the Chicago White Sox logo. You might say that I was crazy, but look for yourself. It's right there. (Hint for sex embed: The bottom curve of the *S* helps form the upper portion of the circle in the letter *e*.) My point here is that the logo conveys several meanings whether intentional or not. Sex and hate work any place, any time, any context.

The urban hip hop culture has embraced the new Chicago White sox logo. Some brothers and sisters dislike the team but love the jacket and hats. Few understand the subliminal draw is the cleverly embedded word *sex*. I quickly point out to them that many of their favorite hip hop or gangsta rap artists wear the logo in their sexually explicit videos. Coincidence?

INTRODUCTION

Let's take another example, the number 40. What's up with that symbolically? In other words, why *40* ounces and not 42 or 23? Well, on the street, weed is sold in forty ounce bags (mostly on weekends, the traditional party time). Look at these other interesting meanings of 40:

- The 40oz became popular during the term of the 40th President of the United States, Ronald Reagan.
- Laws regarding slavery appeared on the books 40 years after Africans were brought to the English colonies.
- How about that 40 acres and a mule that's still owed us?
- During slavery, Blacks were allotted only 40 hours of sleep per week (40 winks).
- Man comes into wisdom at the age of 40.
- They say that life begins at 40.
- A baby grows in a woman's womb for 40 weeks.
- Students go to school for 40 weeks a year.
- During the 19*40*s there was a mass migration of Black people from the South to the North. Our people used to say, "I'm catching 40 North."
- The most popular bag of weed sold on the streets is a 40 ounce.
- Black people in America have suffered 400 (40 x 10) years of oppression.
- A person is paid for every 40 hours worked.
- During slavery times, 40 African male slaves could be purchased for 800 (40 x 20) gallons of rum.
- The Supreme Court goes public 40 days a year.
- The maximum number of lashes given to a slave was 40. The white slave master took this rule from Deuteronomy 25:1-3: "If a man is guilty of a crime, and the penalty is a beating, the judge shall command him to lie down and be beaten in his presence with up to *forty* stripes in proportion to the seriousness of the crime; but no more than *forty*

stripes may be given lest the punishment seem too severe, and your brother be degraded in your eyes."

Biblically, the number 40 symbolizes trials, probation, and bondage:

- Satan tempted Christ 40 days and 40 nights. (Luke 4:2)
- Moses stayed on the mountain 40 days and 40 nights. (Exodus 24:18)
- Christ began to appear to his apostles 40 days after the crucifixion. (Acts 1:3)
- The flood of Noah's time lasted 40 days and 40 nights. (Genesis 7:12, 17, and 8:6)
- The Bible says it took 40 days to embalm the dead. (Genesis 50:3)
- Israel stayed in the wilderness 40 years. (Exodus 16:35, Numbers 14:34)
- The judgement suffered by Egypt lasted 40 years. (Ezekiel 29:11)
- The reigns of both David and Solomon were 40 years. (1 King 2:11 and 11:42)
- The Christian fasting season of Lent lasts 40 days.

The use of subliminal messages in advertising is illegal and unethical, but advertisers get away with the practice because technology makes detection very difficult. As a result, people find it difficult to believe that companies spend major dollars to embed secret messages in TV commercials, print ads, on alcohol billboards and labels–even on direct mail flyers that come to your home. You may not believe it, but that will not change the facts. This is war, and Corporate America does not play fair. Anything goes.

It is our sincere belief that awareness of the many subliminal techniques used to market alcohol to youth and the family will help to counter the effectiveness of these manipu-

INTRODUCTION

lative malt liquor ad campaigns.

I won't apologize for our findings and interpretations. After all, I'm not the one who decided to place a silhouette of a Christ-like image on a bottle of malt liquor or, via symbolism, crucify and burn Christ in effigy in a national TV commercial. I won't apologize for exposing the Greek encryptions of the words *Jesus, Christ, Savior, God,* and *Son* that were written on a label of a very popular malt liquor.

Message to the Alcohol Marketers (and the Devil)

We here on the grassroots frontline level are fully aware of the psychological and financial tactics being used against our people. Our common sense is no longer lying dormant in a pool of miseducation. We have and will continue to decode every image, symbol, slogan, and ad campaign targeted to young people of color.

We have compiled a long list of those Black organizations and Black events the alcohol industry controls. We are watching those so-called Black leaders who continue to take blood money–the whores, the pimps who are conspiring and lying. Everyone caught with their pants down and dress up screwing over the Black community will be charged with the attempted murder of the Black family.

At our workshops people always ask, "Coach, aren't you afraid for your life? These people in the alcohol industry are very powerful. They're not going to stand by and let you hurt their business. Their spin doctors are going to eat you alive." I usually respond by saying this: "A man must stand for something or he will bend to anything." I believe that God has a plan and I am being used to help accomplish it. What will be will be. I walk with God, so I fear no man's evil.

0269141

W A N T E D

NAME: Zima Clear Malt

OCCUPATION: Assassin

ALIASES: Z, Play Water, Z-Up, White Water, Gold Water

DISTINGUISHING MARKS: Grooves around the waist; long, thin phallic bottleneck; clear liquid

NATIONALITY: American (Eastern European ancestry)

VITAL STATISTICS: ht-12", wt-22oz; 4.6% by volume.

AFFILIATIONS: Adolph Coors Brewing Company. Suspect funds right-wing extremist organizations, some with ties to the Ku Klux Klan.

M.O.: Suspect claims to be a unique alcohol beverage. Tells its victims that it is "zomething different." Targets Generation X.

LAST SEEN: Suspect was last seen partying with middle school, high school, and college students. Was even seen in some classrooms disguised as water.

CAUTION: Suspect is armed, dangerous, and financially and politically powerful.

1. Zomething Fishy

...probably the greatest favor that anybody ever did you [sic], was to drag your ancestors over here in chains, and I mean it.[14]
–William K. Coors, Adolph Coors Brewing Company, to a group of minority business owners

I recently heard a story that just about broke my heart. Seems some young getover artists have been pulling a fast one on their unsuspecting teachers. These kids come into class, books in one hand, a styrofoam cup in the other. In that styrofoam cup is a clear liquid that looks just like water, except it's not water. It's zomething different from water. It's Zima. Kids getting smashed right in front of their teachers' faces. If you think it's funny, you'd better ask somebody. *The Augusta Chronicle* reports that "Kids love Zima." It tastes a lot like Sprite.[15]

Question: Who's really getting over here? The kids who get so drunk that they can't function in school or in life, or the Coors Company, who've just been made a little richer?

Maybe the real winners are the right-wing racist organizations that receive hundreds of thousands of dollars in Coors money. For example, the notorious Heritage Foundation re-

Hmmm.....
Check out this PILLOW TALK *between Coors and Black leadership:*

*"This **incentive covenant entered into as of this 18 day of September, 1984 between Adolph Coors Company...and a National Black Economic Development Coalition...including the National Association for the Advancement of Colored People (NAACP), and People United to Save Humanity (PUSH)** for the mutual benefit of Black America and Coors...*

*"Whereas, the parties agree that the goals reflected in the following agreement recognize that as of February 1984 **Black people represented approximately 6 percent of Coors annual volume in its current market area.** Future determination of Black con-*

sumer market will be determined jointly by the monitoring committee;

"Whereas, understanding the share of Coors business within the Black marketplace is very important if Coors is to return a fair share of its income to the Black community, Coors agrees to endeavor to obtain a 10% level of its annual volume and potential from the Black community. Future figures will reflect market position, goodwill and corporate social responsibility. If market position increases or decreases the commitment will be commensurate but not below 8% through 1985. As the benefits of this covenant are realized and Coors' business increases with the Black community the percentage will go beyond 10%;...

"Coors recognizes that advertising including agency, production, and media placement dollars are important both to Black-owned and Black-formatted media as well as to Coors in expanding its marketing opportunities with the Black community.

"Coors agrees to a total budget of $8.8 million for the initial year of this agreement. The expenditures will vary on a regional basis.

ceived $250,000 from Coors in 1973, their first year of operation.[16] The Heritage Foundation, a conservative think tank, influences policy that runs counter to the interests of Black folks. "The Coors family has funded and worked with an interlocking network of groups which have harbored racists, racial eugenicists, a former Ku Klux Klan leader, even a convicted Nazi collaborator."[17]

Not surprisingly, soon after William Coors made his infamous comment to the Black business group in 1984, Coors money started creeping into the Black community through various organizations and banks.[18] *We're* allowing Coors to use our community projects and institutions to launder their dirty money. Brothers & Sisters, don't be fooled into thinking that Coors loves Black folks. Coors wants your money, plain and simple, and they've spent a lot of advertising dollars to get it.

. . .

Zima Clear Malt was launched during the summer of 1994 on national TV with a commercial called "The Barbecue." At first glance, the commercial looks like a typical barbecue party-food cooking on the grill, drinking, and partying males and females. But, Zima spelled backwards is *amiz*, as in something is amiss (wrong). There's a lot more going on in this commercial than barbecue sauce.

When the spot begins, we see a group of seven Generation X adults, four males, three females, about 19 to 24 years old. They appear to be having a good time. The barbecue party is on a rooftop overlooking what looks like the San Francisco Bay area. A cross-styled antenna appears on the roof in the background. The pace is fast. Images jump quickly from one to the next. This is an old advertising trick, the point of which is to so overload and fatigue your brain that you slip into a hypnotic state. You go cross-eyed. Your subconscious mind then becomes more receptive to subliminal messages and symbols. With symbols being the language of the subconscious, the Zima commercial is most powerful at this level of awareness.

The commercial goes something like this:
1. A big fish, whole head in tact, is offered upside down to the male cook. A large incision runs down the center of the belly of the fish.
2. The male cook puts the big fish on the hot grill.
3. One of the males standing directly in front of the fish is wearing a tee-shirt. On the front of the shirt is a large skull.
4. The male cook pours Zima into the mouth of the big fish. The males and females stand around the grill laughing. One woman sits apart from the action on a couch. She says, "I don't eat meat."
5. While Zima is being poured into the big fish's mouth, someone is stabbing its side with a spatula.
6. The Zima spokesman, well dressed in a white suit and panama hat, talks a bit at the end.

Brothers & Sisters, a lot's going on here. The commercial has at least three powerful messages, and all work on the conscious and subconscious levels of the mind at once. They are the crucifixion of Christ, sexual intercourse, and excessive drinking.

Zima and the Crucifixion

Christianity is one of the world's leading religions. It's symbols and themes have been depicted in art and literature for centuries. For millions of people, believers and nonbelievers alike, the symbols and themes have the power to change lives, work miracles, or instill fear and dread. Marketers say

> *"In the implementation schedule Coors will make expenditures to Black-owned and operated businesses in the following areas:*
> *a. **Black-owned magazines***
> *b. **Black-owned newspapers***
> *c. **Electronic media (television and radio)***
> *d. **Print and outdoor advertising***
> *e. **Public Relations and promotion***
> *f. **Community related activities.** . . ."*
> *–The National Incentive Covenant Between Adolph Coors Company and a National Black Economic Development Coalition, September 18, 1984. (The 1990 Coors renewal agreement was signed by the following African American organizations (incomplete list): African Methodist Episcopal Church; NAACP and NAACP-Los Angeles; National Newspaper Publishers Association; and PUSH. Interestingly and to their credit, the U.S. Hispanic Chamber of Commerce did not renew.)*

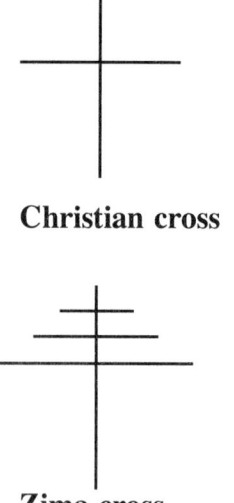

Christian cross

Zima cross

that the Coke logo is the world's most recognizable logo, but that's not true. The cross is, and the Zima commercial uses an antenna to represent it.

The simple cross image speaks worlds about humanity and our physical and spiritual relationship to the universe, to the Creator. The horizontal line symbolizes our relationship to our brothers and sisters, as well as to the earth. The vertical line symbolizes our relationship to the Creator and the universe. Jesus on the cross is the mediator between us and the Creator. In Christianity, the cross was the instrument of Jesus's torture and death. The empty cross symbolizes his resurrection. The Zima spot plays with these life and death themes.

There is a lot of fear associated with the cross, thus it can be used successfully to manipulate and control. During the Inquisition in the Middle Ages, Christian armies marched with the cross leading the way. The Knights Templar wore the cross (rosicrucian) on their uniforms as they went about the world looting and killing in the name of the Lord. It is well known that church missionaries often went to a land to pacify the people first before government armies swooped in like vultures to finish the job. It would not be an exaggeration to say that indigenous peoples across the globe–especially from Africa, Australia, and the Americas–have been crucified on the cross. Lives, land, and cultures have been destroyed.

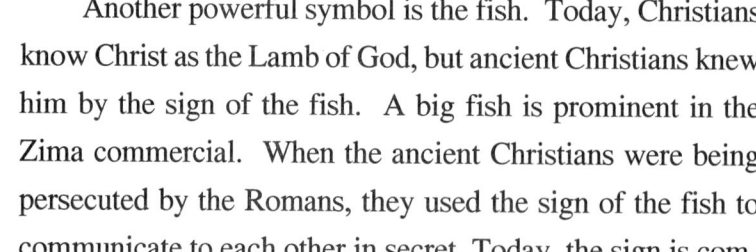

Another powerful symbol is the fish. Today, Christians know Christ as the Lamb of God, but ancient Christians knew him by the sign of the fish. A big fish is prominent in the Zima commercial. When the ancient Christians were being persecuted by the Romans, they used the sign of the fish to communicate to each other in secret. Today, the sign is coming back. Fish can be seen from coast to coast on the back fenders of cars, on walls, in churches, and in the Zima commercial–only Coors is using the sacred symbol to communicate an evil message.

Most of the world's peoples are well acquainted with Christianity's central life, death, and resurrection themes whose centerpiece is Jesus Christ. The Zima commercial banks on the popularity of Christian themes and symbols to sell its product. Let's see how a big fish, a group of males and females, a rooftop, an antenna, a hot grill, some cans of Zima, a spatula, and a picture of a skull work together to re-enact the crucifixion of Christ.

The cross-styled antenna in the background and the rooftop set the stage for this barbecue party. Why a rooftop? While there's nothing highly unusual about barbecuing on a roof, we could argue that most barbecues take place in backyards or parks. We have to assume that the location of the party is important.

In the background is an antenna, which is in the shape of a cross. There is also a hot grill. The fish is being barbecued on the grill. On a very basic level, we decode this as the Christ being burned in effigy. According to the *Reader's Digest Great Encyclopedic Dictionary*, an effigy is "a likeness or a representation." To burn someone in effigy means "to display and **burn**...a crude image of a disliked person." Alcohol is often called spirits, and considering what it does to the body, mind, and spirit, it is an *evil* spirit. Anyone under the influence of an evil spirit cannot love Christ. For them, Christ is a disliked person.

The hot grill (fire) is also classic symbolism for cross. The burning cross was brought into modern day consciousness by the Ku Klux Klan, but in reality, fire has long been associated with the cross and crucifixion. In fact, the words for *cross* in all languages have a common historical form that signifies "light of the Great Fire."[19] Back in olden times when crucifixion was the method most commonly used to kill and torture people, a fire would be lit underneath the feet of the sufferer-not to burn him but to *suffocate* him with smoke to speed up the dying process.[20]

The United Methodist Church in Nashville uses the cross and flame as its official symbol. According to a flyer for the church, "The insignia. . .relates our church to God by way of the second and third persons of the Trinity: the Christ (cross) and the Holy Spirit (flame). . .The flame suggests Pentecost when witnesses saw 'tongues as of fire.' "[21]

Standing next to the grill is a man wearing a tee-shirt that bears a picture of a skull. Mark 15:22 says, "And they bring [Jesus] unto the place Golgotha, which is, being interpreted, The place of a skull." Luke (23:32,33 LV*) and John (19:17 LV) translate the Hebrew word Golgotha as "The Skull." In Matthew 27:33 (LV), Golgotha is translated "Skull Hill."

A hill is high up in the air, and so is a rooftop.

The three elements-the skull on the tee-shirt, the rooftop, and the antenna-combine to recreate the scene of Jesus's crucifixion. Jesus was crucified on a cross that stood on Golgotha, or Skull Hill. In the commercial, a man on a rooftop (Skull Hill) puts a big fish (Christ) on a hot grill (cross).

The word *zima* in several eastern European languages

---

*LV: The Living Bible

means winter, which is, of course, a symbol for death.[22] In Swahili, *zima* means to extinguish, which implies a conscious act of murder.[23] The Zima commercial takes place at a barbecue party, and barbecue sauce is usually a dark red, the color of blood. No doubt about it, this is a death scene, and the Christian symbology makes it a reenactment of the crucifixion of Christ. And there's more.

When the big fish is placed on the hot grill, alcohol from the Zima can is poured into its mouth. Mark 15:23 (LV) says, "Wine drugged with bitter herbs was offered to [Jesus] there, but he refused it." While Jesus was hanging on the cross, he complained of being thirsty. (By the way, the fish (Christ) in the commercial hangs over the side of the hot grill (cross), a sly subliminal of Christ hanging on the cross.) Jesus was offered liquid two times–King James says he was offered vinegar; the Living Bible says wine. The Living Bible is consistent with the purpose of the commercial, i.e., to sell alcohol. Now scripture says that Jesus refused the wine. The big fish (symbolic Christ) in the Zima commercial cannot refuse this wine. Its mouth is forced open and the Zima is poured in. And there's more.

The only book to report the piercing of Jesus's side is John (19:34 LV): "However, one of the soldiers pierced his side with a spear, and blood and water flowed out." Also, "The soldiers did this in fulfillment of the Scripture that says,...'they shall look on him whom they pierced' " (John 19:36,37, LV). In the Zima commercial, while the big fish is on the hot grill, one of the males takes a spatula and appears to pierce the side of the fish.

Three women are featured in the commercial. They represent the group of women who stood with Jesus at the crucifixion. The woman on the couch says "I don't eat meat" (street slang for oral sex), which is something a reformed prostitute like Mary Magdalene might have said. A woman of color wearing dark glasses stands close to the fish on the grill (Christ on the cross). I believe her to be Mary, Jesus's mother. We can't see her eyes, which makes it difficult to read her emotions. She appears to be laughing, but turn the sound completely down and the laugh turns into a scream. Remember, Brothers & Sisters, these subliminal themes are working on your subconscious mind. The action is so fast paced, your conscious mind can't process it all, so it shuts down, and you become hypnotized and receptive. You might *hear* this fake Mary laugh, but you *see* her cry or scream.

And finally, if you still don't believe that this obscene commercial makes a mockery of the crucifixion, check this: near the end of the commercial, after the fish has been barbequed, we see the fish hanging on a wall. This image flashes onto the screen for about two seconds, barely long enough to register in the conscious mind. The big fish is now hanging and mounted, which reinforces the crucifixion theme.

Zima and Sex

> Most of our females who become abusers of alcohol, or are alcoholic, will be sexually abused while under the influence of alcohol. Whether she is 12 years old or 50 years old, being sexually abused for these sisters is almost a guarantee for them. We know that other consequences include unplanned pregnancies and STD's (including HIV).[24]
> –Beverly Hall-Ogletree

Sex sells, and the Zima commercial is thick with sexual themes and symbols.

Prior to being put on the grill, the big fish is raw. Raw fish is cold. The big cold fish is offered to the cook in an upside down position. A large incision runs down the center of the cold fish's belly. The cook takes the cold fish and lays it on the hot grill. Then he takes a can of Zima Clear Malt and pours it directly into the mouth of the cold fish as the males and females stand around the hot grill laughing.

What kind of pornography is going on here? Fish is symbolism for female genitalia (vagina). The word *zima* is masculine. *Zima* also means winter. Winter is often referred to as an old man in art, i.e., "old man winter." An old man has a penis. A penis holds three primary fluids: red blood, yellow urine, and white (or clear) semen. Bottles, cans, and the alcohol itself are all phallic symbols–bottles and cans in their shapes, alcohol in its color (yellow or clear). Zima's slogan is "zomething different." Clearly, the color of this malt is different. Unlike the traditional golden yellow urine color of most malts, Zima is clear, like semen. Think a moment about the act of pouring Zima (read: semen) into the mouth of a fish (read: vagina). This suggests a sex act, because the Zima (semen) is being poured (forced?) into the *mouth* of the fish. Now, a cold fish is cultural symbolism for a woman who refuses to have sex. The act of pouring Zima (semen) on a cold fish (woman) suggests rape. In this light, the women's comment, "I don't eat meat" and "give me more, give me more" adds yet another symbolic layer to the scene. Say Zima, say semen. They even rhyme.

Now remember, this is all happening at a barbecue party. Sex is a major theme at most adult parties, and any woman who refuses to have sex is called a "cold fish." The commercial suggests a way to heat up a cold fish (frigid, unresponsive female). Give her Zima! The hot grill upon which the big fish lies is a symbolic hot bed. When the man pierces the side of the fish, the tangle of sex and violence makes for a potent symbol. There is a feeling of rape here. The fact that the males are standing around laughing suggests a gang rape.

"It doesn't make sense to pour alcohol into the mouth of the fish considering the fish has a big incision in its belly. The alcohol would simply flow out. This to me means that the body of the fish (Christ) is refusing to accept the alcohol."
–Pam Thomas, Dayton, OH

Zima Gold is one of the latest Coors products out on the market, and judging from the print ads, this dark colored brew is being targeted to dark people (people of color). The ad shows some sweaty Black male basketball players relaxing on a bench, enjoying a fresh bottle of Zima Gold after a hard game. The slogan reads, "You better ask somebody." Perhaps the brothers should ask somebody if strenuous workouts like basketball and alcohol mix. Such a mixture can lead to heart failure. This ad helps to fuel what Richard Pryor calls the "supernigger mentality," i.e., "I'm invincible!" Brothers, don't believe the hype. Alcoholism is the number one killer disease in Black America.

Remember the woman of color who stands closest to the fish? With the sound up she is laughing, which suggests her cooperation in this act of rape. In fact, she likes it. Rapists often say their victims were "asking for it," even enjoyed it. With the sound completely down, however, this woman of color looks like she's screaming, which reinforces the general helplessness of women when being raped. And there's more.

The large incision that runs down the belly of the big cold fish presents troublesome issues for women. When you gut a fish (woman), you remove all of its eggs/ovaries/uterus. An incision running down the belly of a female could suggest that she has gone through any number of operations–a hysterectomy, for example. A full hysterectomy brings on menopause in a woman. Her menses stop, and she is no longer able to have children. Women's traditional role in society has been to bear children. When she is no longer able to fulfill this function, she is put out to pasture. She is no longer desirable among men. Menopause is a difficult transitional time for many women. Many feel the essence of their femininity has died along with their biological ability to bear children. The cold fish as woman *cannot* bear children because she is biologically unable or *will not* because she does not want to have sex.

An incision running down the fish's belly could also suggest another equally disturbing operation: an abortion. Abortions are, of course, about death, or, as the swahili definition of *zima* states, an extinguishing of life.

Clearly, this commercial is obscene. In fact, the commercial was pulled from the air and revised because the "f" word could be heard during the latter part of the commercial.[25]

ZIMA CLEAR MALT

Zima and Excessive Drinking

We often say that alcoholics and other heavy drinkers "drink like fish." When Zima is poured into the mouth of the fish, the sly subliminal command to the target audience is "drink like a fish." In fact, from the perspective of the fish, it appears to be guzzling the alcohol. Guzzling is a rite of passage for many teens and young adults pledging gang or fraternity membership. Alcohol is also a big part of young adult partying; anyone who refuses to "drink like a fish" is made to feel like a "fish out of water," i.e., a killjoy.

While Coors makes excessive drinking look fun, nothing could be further from the truth. The commercial doesn't tell you that alcohol irritates and inflames the esophagus, causing esophagitis, a painful condition that interferes with swallowing. An inflamed esophagus will produce burning pain, nausea, vomiting, and loss of appetite which contributes to extreme weight loss and nutritional problems.[26]

0000-45

WANTED

NAME: Colt 45

OCCUPATION: Assassin

ALIASES: 6 Shooter, Gun Slinger, The Great Equalizer, Little Red Pony; Premium Colt 45, Colt 45 Double Deuce, Colt 45 PowerMaster

DISTINGUISHING MARKS: Horseshoe, a bucking red horse, the number 45

VITAL STATISTICS: ht-12", wt-22, 32, 40, 45, and 64oz. Loaded with sugar. Up to 7.1% by volume. Cheap- about $1.39 per 40oz.

CRIMES: Corrupts minds, destroys families and communities. Contributes to school truancy, high dropout rates, violence, promiscuity, unwanted pregnancies, STDs (e.g., AIDS), birth defects, rapes, homicides, and self hatred. Causes spiritual deprivation, emotionalism, sexism, denialism, egotism, poorism, and alcoholism.

AFFILIATIONS: G. Heileman Brewing Co. Suspect is an active member of the multimillion dollar cartel that launders money through African American events and organizations in exchange for the silience of Black leadership.

M.O.: Suspect claims to "work every time." Gains trust through innocent beer disguise. Black entertainers hired to front for G. Heileman. Targets Black people from 6 to 60.

LAST SEEN: Suspect was last seen partying with Black folks on street corners, playgrounds, parks, and college campuses. Sighted in several liquor stores in ghettoes across America. Eye witnesses report seeing suspect in Black exploitation films, TV commercials, Black magazines, and billboards in Black neighborhoods.

CAUTION: Suspect is armed with sugar and a higher alcohol content than the law should allow. Is extremely dangerous while in the company of firearms. If seen, stay clear!

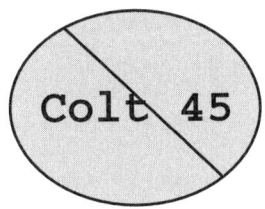

2. The 40oz Harness

Before the Colt 45 PowerMaster upstrength brand had a chance to walk out the door of the G. Heileman Brewing Co., the community activitists were on the case. Why? Because PowerMaster was being marketed to Black people. The same PowerMaster that boasted 7.1% alcohol by volume–the most potent malt liquor on the market at the time. In the early '90s, Chicago priests George Clements and Michael Pfleger, New York minister Calvin Butts, the Marin Institute, and others, slammed G. Heileman Brewing Co. for naming its now defunct malt liquor "PowerMaster." The alcohol industry's own code states that advertising should not carry any hint of strength. Although G. Heileman denied that the name PowerMaster was meant to indicate strength, truth is, the name sounds like it belongs to some superhero, like one of the Power Rangers or Masters of the Universe.

G. Heileman got cold busted and was forced to surrender the PowerMaster name. So the company introduced a "new" brew, "Colt 45 Premium." Same drink, same tall Black can. According to Makani Themba, public policy specialist, Marin Institute,

> The term Premium targets inner city Black men. Furthermore, G. Heileman's tagline slogan "be a

"We know they're targeting the black community and we're sick of it. This [PowerMaster] is a drug and we're calling them [G. Heileman] drug dealers."
–Deborah Fair, President, Michigan Black Alcoholism Council, Los Angeles Times, June 25, 1991

"The marketing plans for PowerMaster are socially irresponsible."
–Surgeon General Antonia Novello, Washington Post, June 26, 1991

"PowerMaster is little more than a magic bullet for bankrupt brewer G. Heileman that is being aimed at the black community."
–Rep. John Conyers (June 26, 1991 press release)

player" links the drink to inner city connotations of power and sexual success. In street slang, a player is a hip ladies man. In every city you find a gang that refers to themselves as players.[27]

It's clear that Colt 45 PowerMaster/Premium is designed for and targeted to Black folks in the hood. This drink has put many a Black person under subjection. It has made them slaves, blind disciples, sheep-like followers, addicted.

What's in a name? A name reflects a value or standard. It conveys an expectation. It speaks intention and recalls historical events, memories, and emotions. Let's assume that G. Heileman's first name choice comes closest to the company's philosophical idea for the drink. Every product, from gym shoes to jewelry to malt liquor, has a purpose beyond profit that reflects the company's ideology. When DeBeers says "A diamond is forever," we know that the company intends to maintain its stronghold in South Africa and the global diamond market for a long time. Thus decoding the name PowerMaster would be a more meaningful exercise than decoding the name Premium. And since the drink is targeted to Black folks, we must decode from an African-centered perspective. Remember, G. Heileman denies that the name connotes strength, and we believe the company would never deceive the Black community or break their own industry's rules and regs. So what does the name mean?

Webster says that

- *Power* is "the controlling influence, a person of great influence or authority. An indicator of one's might and strength. A person who owns wealth–mass property or land–has power."
- A *master* is "a man in control or authority of others completely."

G. Heileman says the original name did not speak to alcohol strength. Webster's definitions suggest that they might have been telling us the truth. Taking the definitions one tiny step forward, we can say that a power master is really a slave master. Oh no? Oh yes! Review the definitions. Notice that "power" and "master" have control and authority in common. A slave master is a person who owns slaves, a person who has complete controlling authority of another person.

Brothers & Sisters, could it be that PowerMaster/Premium's job is to make you a slave? It wouldn't be the first time alcohol was used to keep the slave master in power.

Let's work the signs. The following sections will analyze each of the elements on the Colt 45 label, as well as various ad campaigns.

The Horse

I strongly suggest that you read *From the Browder File* by Anthony Browder. In it the brother says that "Signs are wonderful things. They tell us where we are, where we are going, or where we have been. When we properly interpret signs, they can yield some valuable information."[28] For years Billy Dee Williams has been telling Black folks to "Read the signs." Why would G. Heileman want us to read the signs? What signs? The only signs they've given us to read are the ones on the label, billboards, posters, and TV commercials.

The first sign, or symbol, that appears on the label is the word "Colt." Webster defines a colt as a "*Male*, the *young male* of *any species*, the young male *horse*, a young *inexperienced* person, i.e., a *young ass*." (My emphasis.) We know that G. Heileman uses Black stars like Billy Dee Williams to sell Colt 45 to Black folks. According to a 1984 article in *Advertising Age* magazine, Blacks were buying 75% of all

Not every Black star has sold out to dirty money. Stand-up comedian and Hanging With Mr. Cooper *star Mark Curry didn't sell out. "They already asked me to do a Colt 45 commercial," says Curry. "Man, I can't do a Colt 45 commercial. Colt 45 is directed at the black people; it's only sold in black areas and it's a poison."[30]*

Colt 45 malt liquor sold.[29] Today in 1995, a modest estimate would be 80-90%. Question: Which segment of the Black community is Colt 45 being marketed to? Webster gave us the answer: Young Black males! Compare the Black male experience to Webster's definition.

Webster	**Black Men**
male	Young Black men are *male*.
young male of *any species*	Young Black men are called an endangered *species*.
the young male *horse*	The sporting world refers to Black men, the super athletes, as *horses*. Also, during slavery, Black men were referred to as work *horses*.
a young *inexperienced* person	Although Black men are experienced in the ways of the street, they are *inexperienced* and naive to the ways of Corporate America. They regularly get played. They are often convinced to buy things they don't need or will do them physical and/or spiritual harm.
a *young ass*	Whenever Black men become intoxicated they behave like *young asses*.

Horse Play

In the summer of 1992, TV viewers were assaulted by a commercial that depicted a Black man and a Black woman breeding. Don't believe it? In the commercial, the Black male and Black female both wear black horse costumes. The Black male dances erotically between the legs of the Black female. The male is like a stud horse–a breeder, that is. This is an obvious appeal to the subconscious of Black males. Also,

42

with the horse head masks, the actors look like centaurs, the half-man, half-horse monster race in Greek mythology. Clearly, the Colt 45 logo is more than just horse play.

Within an historical context, such symbolism has far reaching implications for African Americans. On the surface, the depiction of a Black male as a horse simply reinforces the colt-Black male connection. Taking it to a deeper level, during slavery, the Black male was called by many of the same names that the slave master called his horse (his pride and joy).

Horse Type	**Meaning**
stud horse/stallion	sex breeder
thoroughbred	fighter (super athlete)
colt or young buck	young male slave
work horse or mule	laborer, mentally slow
jackass	a fool

The slave master would often have his horses and slaves race for entertainment. The breaking and breeding of slaves was cruelly patterned after the breaking and breeding of wild horses. These methods of dehumanization became known as the William Lynch Method. William Lynch was a slave owner from the West Indies who was famous for his so-called fool proof "slave breaking and breeding process," also known as "the cardinal principles for making a negro." Lynch stated, "The breaking and breeding process is the same for both the horse and the nigger. Only slightly varying in degrees."[31]

The Colt 45 mascot is a bucking red horse. A horse bucks in fear or when there is an attempt to tame it. Let's look at both angles. Black men are in a constant state of stress and fear. They lead all other groups in stress-related diseases, such as hypertension. Although the code on the street is "never let them see you sweat" or "chill" or "be cool," beneath the Black man's cool facade is a deep fear of failure- in bed, in school, on the job, in the job search, in relationships with women, as fathers, as men to be respected in a white male dominated society.

If, on the other hand, the horse on the label is being broken in, we could say that it is in a state of rebellion. Strength and skill would be required to tame this horse (Black male)- which would account for the super high alcohol content and countless millions of dollars that have been poured into marketing research and advertising campaigns. It takes a lot of dope in a bottle and brainwashing to tame the strong, rebellious Black male, the signs on the label seem to say.

The Red Horse

The horse on the Colt 45 logo is red. Question: Why a *red* horse? The slogan tells us to "Read the signs." One of the best sources for sign reading is the Bible, specifically, the book of Revelation.

There are 22 chapters in Revelation, so trying to find answers there would be like searching for a drop in a 40oz bottle. We need some kind of clue, so let's start with the label. The only number on the label is 45. Since there is no 45th chapter in Revelation, split 45 to read 4 and 5. Now read Revelation 4:5. You'll note the scripture isn't consistent with the symbols and images used in the ad campaign, so let's try again. List all the numbers that appear before the numbers 4 and 5. Your answer should be 0 1 2 3. Now add the numbers: 0+1+2+3=6. Six is your chapter. Double check your answer by asking yourself what number follows 4 and 5. Answer: 6.

What verse do they want us to read first? Once again, split 45 to read 4 and 5. The first number is 4, i.e., verse 4.

And it reads,

> And there went out another *horse that was red*: and *power* was given to him that sat thereon to take peace from the earth and that they should kill one another: and there was given unto him a great sword.

The rider of the red horse is given authority to remove peace from the earth. The absence of peace is war. Even though Revelation 6:4 clearly interprets itself, further proof is given in the parallel prophecy of Matthew 24:6-7:

> And ye shall hear of *wars* and *rumors of wars*; see that ye be not troubled: for all these things must come to pass, but the end is not yet. For *nation shall rise against nation*, and kingdom against kingdom....

Brothers & Sisters, is it a coincidence that the logo for Colt 45 PowerMaster/Premium label is a red horse? The one in Revelation who sat on the red horse is literally a power master with the power to kill. Note that the verse speaks of violence and death. According to *U.S. News & World Report*, "Over the last several decades, roughly half of all Black homicide victims and perpetrators had been drinking at the time of the crime. According to the FBI 1992 statistics, 50% of all homicides in America involved alcohol."[32]

The red horse icon has been embraced by street gangs. Street gangs call themselves nations. These nations often go to war over kingdoms (turf). Prior to battle, there are always "rumors of wars" in the hood.

Why a red horse as a logo? According to the Bible, the red horse is the omen for war. When we say we see red, that means we're angry enough to hurt somebody. These war symbols of aggression and violence make for a very dangerous subliminal message. Add to the fact that Colt 45 is the name of a gun used in the wars of America, and we're reminded that the Black community is at war with itself and somebody else!

Note that the red horse on the label appears to be bucking very violently. It looks like the red horse has suffered some kind of trauma. In many cultures, e.g., indigenous West African and North American tribes, the warriors would dress their horses in paint before going to fight. The war horse was trained to return to the village after each battle, with or without its warrior. Often the war horse would return covered from head to hoof in the blood of its fallen warrior. The horse covered in red blood would serve as an omen to the villagers that war was upon them.[33] The bloody (red) horse on the Colt 45 label is symbolically warning the Black community that *war is upon us*.

The Horseshoe

Since we know there is no such thing as a red horse in the physical world, let's wash the blood away (symbolically speaking) to reveal the horse's true color. Not only does G. Heileman tell us to "Read the signs," the company tells us that "It works every time, no doubt!" Who and what are working every time, no doubt? Brothers & Sisters, there is only one other horse that has the power to work every time, and that horse appears in Revelation 6:8–

> And I looked, and behold a *pale* horse: and his name that sat on him was *Death*, and Hell followed with him. And *power* was given unto them over the fourth part of the earth, to kill with *sword*, and with hunger, and with death, and with the beasts of the earth.

46

Another coincidence? Don't believe your eyes? How did the signs add up to Revelation 6:8? Let's double check the clues on the label. The symbol following the word Colt is a horseshoe (U). On a superficial level, the horseshoe resembles a magnet, which symbolizes the magnetic attraction advertisers and product developers hope to get you to feel for their malt liquor. It gets deeper. "Western mythology says a horseshoe that appears in an upright position means good luck."[34] When the bottle is straight up, the horseshoe is upright, thus signifying good luck. The opposite must also be true. When you turn the bottle or can upside down to drink, the horseshoe connotes bad luck. Alcohol leads to bad breath, bad looks, bad body order, bad timing, and all that's bad luck! And it gets deeper than that.

You said you wanted to know how I got Revelation 6:8. Every time you turn the Colt 45 container upside down to drink, the horseshoe (U) is transformed into the omega (Ω) symbol. Omega has a numeric value of 800 and a symbolic value of 24 within the original* Greek alphabet.[35] (More on 800 in chapter 3.) Now take 24 and split it to read 2 and 4. Add them, then multiply them.

$$2 + 4 = 6 \text{ (That's chapter 6.)}$$
$$2 \times 4 = 8 \text{ (That's verse 8.)}$$

The upside down horseshoe, or omega sign, symbolizes "the end" or "the last." ("I am Alpha and Omega, the begin-

---

*According to E.W. Bullinger, *Number In Scripture*, there were 24 letters originally in the Greek alphabet. Omega (Ω) is the 24th and last letter. However, at some point the Greeks decided that they wanted 27 letters like the Hebrew's had in their alphabet, so they added three more letters: stigma (6), koppa (90), and sampsi (900). For our decoding purposes, we'll stick to the original.

ning and the ending," God says in Revelation 1:*8*. Note that verse 8 equates to omega's numeric value of 800 - 8+0+0=8.) Brothers & Sisters, the end, the last of what? G. Heileman's signs and slogan lead us to the book of Revelation, which deals with the last days and the end of time. Another coincidence?

Which letter in the English alphabet, and what word in the English language, does the upright horseshoe most closely resemble?

$$\cup = U$$
$$\cup = You!$$

You what?

U in the English alphabet is the 21st letter; thus it has a numeric value of 21.
Ω in the Greek alphabet is the 24th letter; thus it has a symbolic value of 24.

$$21 + 24 = 45!$$

Regardless-whether the container is upright ($\cup$) or upside down (Ω), the numeric value + the symbolic value adds up to 45.

The slogan for Colt 45 PowerMaster is **"bold, not harsh."** The treatment dished out by the riders of the red and pale horses will be bold and harsh. Many biblical scholars agree that the rider of the white horse is a counterfeiter, a fake, a liar, the Antichrist himself. The alcohol industry rides through the Black community pretending to be friendly and generous in its philanthropy to Black causes while pouring out alcoholism and death. Could it be that the Antichrist is not an individual but an industry, i.e., the alcohol industry?

One final point about the horseshoe. Everytime you place a container of Colt 45 malt liquor horizontally to your lips, the horseshoe on the label symbolically takes on the appearance of a bow, as mentioned in Revelation 6:2:

> And I saw, and behold a white horse: and he that sat on him had a *bow*; and a crown was given unto him: and he went forth conquering and to conquer.

In the Old Testament, the bow is always a sign of military power.

> He maketh wars to cease unto the end of the earth; he breaketh the *bow*, and cutteth the spear in sunder; he burneth the chariot in the fire.
> –Psalms 46:9

> And it shall come to pass at that day, that I will break the *bow* of Israel in the valley of Jezreel.
> –Hosea 1:5

> Because the spoiler is come upon her, even upon Babylon, and her mighty men are taken, every one of their *bows* is broken: for the Lord God of recompence shall surely requite. –Jeremiah 51:56

The Jive of No. 45

Following the horseshoe symbol on the label is the number 45. Earlier we used 45 to help us find the chapter and verses to decode the signs on the Colt 45 label. We've established that the colt represents the Black male and the horseshoe symbolizes "you." Put them together and we have the beginning of a secret message that reads "Black male [Colt], you [U]." Obviously, "Black male, you" is an incomplete phrase. We're going to go deeper into 45 using the system of alphanumerology to finish the sentence. In alphanumerology, each letter in the English alphabet has a number based on its position (e.g., A=1, B=2, etc.).

Message 'n a Bottle

Hmmm...

The following Black men are noted on a Colt 45 money laundering "give back to the community" ad: Congressman Adam Clayton Powell, Dr. Charles Drew, Dr. Martin Luther King, General Colin Powell, Coach Art Shell, Morgan Garrett, Governor L. Douglas Wilder, George Washinton Carver, Elijah McCoy, Senator Edward Brooke, Ralph Bunche, and Matthew Henson. Would these men, dead or alive, appreciate their names being used to sell alcohol to the Black community?

This is the same system used by the U.S. Treasury Department. Got some cash in your pocket? Take out a bill in any denomination. Examine the face side, then look to the left of the picture. A letter appears inside of a circular graphic. To the immediate left of that letter is a number. That number is the numeric equivalent of the letter. You will find this number in each corner on the face side of the bill.

Now let's further decode the number 45. There is no 45th letter in the alphabet, so split 45 to read 4 and 5. Add: 4 + 5 = 9. All multiple digit numbers have hidden positive (+) values as well as negative (-) values. Let's keep it simple and figure the positive value of 45: 4 + 5 = 9. Using the system of alphanumerology, figure the letter equivalent of the numbers:

4	5	9
D	E	I

Now, unscramble these letters to find the mystery word. It is DIE.

Colt = Black male
U/☾ = you
45 = die

Colt + U/☾ + 45 = ***Black male you die!***

Obviously we have failed to read the signs. Maybe we've been too intoxicated as a people to read them. And isn't it interesting that the symbols were put in the right order to read "Black male you die." Coincidence?

Brothers & Sisters, this message, this *curse*, is consistent with the other symbols on the label and the deadly effects malt liquor has on mind, body, and soul. The red horse is an omen for war and is associated with death as mentioned in Revelation 6:4. Colt 45 is the name of a powerful handgun which is also connected to death. The alcohol industry is killing our young people spiritually, psychologically, and physically. Does G. Heileman have a death wish for the Black community? Harry X Davidson suggests the following:

> Suppose you came home one day and found bullet holes in your windows. Later, that night, you found a rattlesnake in your bed. The next morning you discovered a tarantula in your bath tub and rat poison in the sugar. What if you then went to your car and discovered a bomb under the hood, and turned around to face a semi-truck bearing down on you? How long would it take you to realize that somebody was trying to kill you?[36]

Hmmm . . .
The words "power" and "master" are both commonly used to refer to God.

Let's return to the number 45. Remember, the letters D I E? Put them back in their original order:

4	5	9
D	E	I

Brothers & Sisters, you're not going to believe this, so hold on to your hats. The word *God* is also written in code on the label! Yes! No wonder G. Heileman fought so hard to keep their PowerMaster name. Look at these **powerful** and **masterful** *dei* words.

Dei-ty: a god or goddess
Dei-fy: to treat as a god; to make a god of
Dei-cide: the killer or killing of a god
Dei-sm: a belief in God

Most alcoholics love their drink more than they love God. Many people treat alcohol as a god. There's an African proverb that goes, "Beware of the false prophet, for his wine will betray you." Both the Bible and the Holy Qur'an warns us about alcohol. This evil spirit undermines your ability to commune with God. Before you know it, you are spiritually disconnected, and not even AT&T will be able to reach out and touch God for you!

PowerMaster/Premium kills the God inside of you. Wake up, Black America and read the signs. Someone is trying to kill your ability to love yourself, your God, and your people.

The Six Shooting Colt or Cult?

I know what you're thinking. "This is some good science fiction. No way can this brother expect us to believe all this! Too crazy." But as brother Dick Gregory has said, "Information is power!" G. Heileman and its competitors would mislead and miseducate our spirits and minds, push us down the road of fratricide and genocide, all the while pretending to be innocent. They must believe that poor communities of color will never defeat their beast mentality. But we can. We got no choice.

Question: What kind of group slanders, assumes the name of, and mocks the worship of God? Answer: A cult. Cultists seek the power to master and manipulate the forces of the universe. Satanic cults and occultists are under 666, the mark of the beast. That 666 pops up in the most amazing places. Colt 45 isn't only a malt liquor; it's the name of a very powerful handgun that holds 6 bullets, causes 6 wounds that can cause 6 deaths. Now compare and connect the similarities:

	6	6	6
C<u>U</u>L T	6	6	6
C<u>O</u>L T	6	6	6
	BULLETS	WOUNDS	DEATHS

It gets deeper.

The cult's number:	6 +	6 +	6	= 18
Colt 45:	6 +	6 +	6	= 18
	BULLETS	WOUNDS	DEATHS	
Colt 45 malt liquor:		4 + 5		= 9

45!

(as in Colt 45)

Another coincidence? Every time you drink Colt 45 you symbolically blow your brains out! And it gets deeper. Check this out:

SECRET CODE SCRAMBLED

4⁺	5	⁼9
D	E	I
D	I	E

SECRET CODE DECODED

4⁺	5	⁼9
G	O	D
D	I	E

The decoded message is "God die," which is typical obscene cult-like language. This isn't a sophisticated message, but it doesn't have to be. You get the point. When you drink, the alcohol undermines your ability to commune with God. Satan, on the other hand, will become loud and clear. He'll tell you to rape your sister. He'll tell you to kill your brother. He'll tell you to open your legs to a train of men. The spirit of alcohol, especially upstrength malt liquor, is evil. Most fights and homicides occur after a few drinks. God never dies, but your ability to worship God in spirit and in truth can be severely damaged.

> O people of the scripture! Why confound ye the truth with falsehood and knowingly conceal the truth? –Holy Qur'an (Surah 3:71)

Someone somewhere knows the truth about all these "coincidences."

The Pyramid

Brothers & Sisters, we won't leave any stone unturned. Colt 45 Double Deuce malt liquor, the 22oz starter kit, features an upside down pyramid. On the inside of the pyramid is the number "22." Here is yet another clue to the beast. Once again, the book of Revelation will be our guide. To find the relevant chapter and verse, split 45 to read 4 and 5. Add: $4 + 5 = 9$. When you drink from the bottle of Double Deuce, thus tilting the bottle up, the downward pyramid points upward. The upward pyramid is also the Greek symbol delta (Δ), which has a numeric value of 4. Add: $4 + 5 + 4[\Delta] = 13$. Thirteen is your chapter. To find the verse, take the original equation ($4 + 5 = 9$) and add all the numbers together: $4 + 5 + 9 = 18$. Eighteen is your verse. Revelation 13:18 reads:

> Here is wisdom. Let him that hath understanding count the number of the beast: for it is the number of a man; and his number is Six hundred threescore and six [666].

The pyramids were built by Africans in Kamit (Egypt) on the continent of Africa. The upright pyramid symbolizes eternal life. A downward pyramid symbolizes death.[37]

On the 22oz bottle, the pyramid points down. The number 22 appears in the lower portion of the pyramid. Just below the number 22, inside the pyramid, appears the abbreviation for ounce, or "oz." When you drink from the bottle, not only does the pyramid turn right side up, the word "oz" becomes "zo," which is a primitive Egyptian radical meaning "life."[38]

Clearly, the symbolism of the pyramid in both its upward and downward positions, combined with the hidden information buried within the marketing campaign for Colt 45 PowerMaster/Premium, leads me to decode the complete label message as follows:

Colt	∪	4 5 9 / D E I / D I E	Δ[4]		This Is Your
Black Male	**You**	**DIE**	**For**		
Bold Not Harsh		Ω[omega] / **Last**	▽ ZO / **Life!**		

> *Black male you die, for this is your bold not harsh last life!*

Alcohol is a life and death situation for the Black community. Brothers & Sisters, think about the following:

For every message is a limit of time, and soon shall ye know it. –Holy Qur'an (Surah 6:67)

The Colt 45 Dry Song

When I first heard the Colt 45 Dry song I was stunned, stupified, amazed, and tripped out. You will be too. They put the truth up in our collective face! Have we been too drunk to notice?

> *Everybody shakin'*
> *to those body sounds*
> *See those bodies bent*
> *Hear them hearts pound*
> *The room is quakin'*
> *See those muscles sweat*
> *All screaming thunder*

Message 'N A Bottle

Hot as it gets
Heat it up
Shake it baby
Dry down
Colt Dry
Turn up the heat
Dry down!

The jingle is rapped hip hop style to a hard driving funky beat, which is appealing to youth. To twist sister Roberta Flack's words, they're killing us loudly with their song. They tell us the truth, yet we keep buying. Has this Colt 45 truly blown our brains out? The following is my line by line read. See if you agree with my interpretation.

Everybody shakin'
Acute body tremblers. Chronic adrenaline shock?
to those body sounds
Your body asking for help. Kidney failure, fatty liver, pancreatitis, esophageal cancer, cirrhosis of the liver, ringing in the ears, impotence (males).
See those bodies bent
Bent is street slang for being high. Bent to the curve. Hangover? Bent over throwing up from alcohol poisoning and spasms?
Hear them hearts pound
Arrhythmia (irregularities of the heart), tachycardia (heart palpitations), hypertension, anxiety.
The room is quakin'
Head spinning, dizzy spells, alcoholic stupor, staggering, vertigo.
See those muscles sweat
Night sweats, chills, muscle spasms, muscular cramps, dehy-

dration, alcoholic neuropathy, hypoglycemia.

All screaming thunder

Alcoholic hallucination, psychosis, phobias, night terrors, temper tantrums.

Hot as it gets

Hot flashes, physiological withdrawal symptoms.

Heat it up

Slang for "heat up the dope," i.e., crack, heroin, or weed.

Dry down

Numbness, dry mouth.

When decoded, the theme song for Colt 45 Dry gives a general description of the detoxification process, a.k.a. DRYING OUT. The song alludes to symptoms related to hypoglycemia, which is a deficiency of sugar in the blood. "In a 1981 survey. . 93 percent of alcoholics tested showed disturbances in blood sugar regulation..."–i.e., low blood sugar.[39]

MUSTANG'S BLACK HORSE

Colt 45's red horse wasn't enough. It was only a matter of time before some brewer (the Pittsburgh Brewing Co.) would create a campaign marrying a black horse to alcohol, then attempt to shove it in the brains (and down the throats) of the Black community. The Mustang malt liquor label depicts a bucking black horse standing on its hind legs. Webster says that a mustang is a "half wild horse; a stray animal, small but tough." Webster does not mention color, but I'm sure mustangs come in all colors–all the colors horses come in, that is. So why a *black* mustang?

> And when he had opened the third seal, I heard the third beast say, Come and see. And I beheld, and lo a **black horse**; and he that sat on him had ***a pair of balances*** in his hand.

Message 'n a Bottle

WHEAT BARLEY

And I heard a voice in the midst of the four beasts say, A measure of *wheat* for a penny, and three measures of *barley* for a penny; and see thou hurt not the oil and the wine. –Revelation 6:5,6

The black horse obviously relates to scripture, but there's an even more amazing connection. Turn the bottle cap downward. You will see a pair of balances, or a scale, weighing wheat and barley, just like it says in Revelation 6:6. This looks like a recipe for making beer. Another coincidence?

The West depicts everything evil, wrong, or underhanded as black. A black horse is symbolically the same as a "dark horse," and we all know that dark horses are losers. In the movies, the bad guy always rides the black horse, wears a black hat, drives a black car. *We* know that black is beautiful, but to understand where the alcohol companies are coming from we've got to look at *black* from their point of view. Brothers & Sisters, if you choose to ride this black horse (alcohol), you are bound to finish last!

05-800

W A N T E D

NAME: Olde English 800

OCCUPATION: Enslaver

ALIASES: 8-Ball, OE, Olde Gold (Midwest), Olde E (East Coast), Olde British

DISTINGUISHING MARKS: Angry looking tiger, fearful looking lynx (with baby teeth), phallic bottle, golden yellow liquid

VITAL STATISTICS: Varies. Has been spotted in the form of 22, 40, and 64oz bottles. Can be dangerous in any size. 6% alcohol by volume. $1.39.

CRIMES: Ethnic cleansing against indigenous peoples throughout the world. Corruption of the community. Insights self-hate and alcoholism. Wages war on the Black family and community.

AFFILIATIONS: Pabst Brewing Company. Has strong ties with gangs and underaged drinkers in the hip hop/rap culture.

M.O.: Uses Black sex symbols and animals in billboards and ads posted at liquor store cash registers. Launders money through rap contests.

LAST SEEN: Your house?

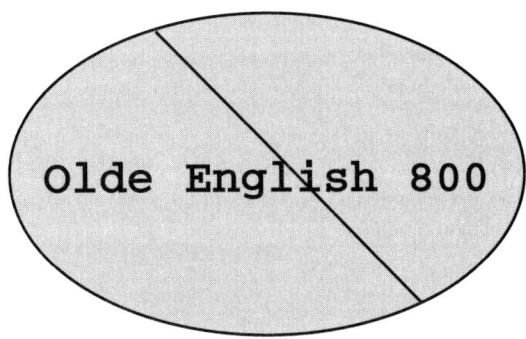

3. The British Are Coming!

The incredible success of Olde English 800 malt liquor suggests to me that far too many Black people are out of touch with the history of the African holocaust (maafa) experience. A marketing brochure for Olde English 800 noted that the product is "brewed for relatively high alcohol content (important to the ethnic market)!"[41] Pabst Brewing Co. has a serious scheme to take our money, our minds, and our lives, and we're falling for it. The more I see Olde English 800 malt liquor in the hands of young Black men and women, the more I'm convinced that the American educational conspiracy to stamp out African memory is in full effect.

Have we forgotten who snatched 150+ million African children, women, and men from the shores of their beautiful land, capitalized on the trade of human cargo, launched a wholesale genocidal attack on the Native Americans, then had the nerve to call themselves landlords, slave masters, and Christian? Lest we forget again, it was the olde English. Although many European countries were involved in the African slave trade, including Portugal, Spain, Holland, France,

"We are a drunken land today and a drunken race–not because of prohibition or the open saloon, but because we have ceased to teach temperance to the young."[40]
–W.E.B. DuBois, 1928

Sweden, and Denmark, it was the olde English who monopolized the unholy trade. When Pabst decided to name their malt liquor after the same people who enslaved our ancestors, it was an "in your face" move against the Black community. And since Pabst chose to promote their high octane malt liquor to communities of color, we have the right, in fact it is our duty, to expose the murderous legacy of the olde English.

For four years my partner Shack and I have crisscrossed this country doing workshops, gathering data, and conducting interviews for this project, and for four years we have witnessed the national epidemic of young people sucking on the slave master's bottles. Some of our lost brothers and sisters show pride by lining the window ledges in their homes with empty 40 and 64oz bottles, displaying them as if they were trophies. Daily viewing of these empty bottles has a powerful effect on the mind. Seeds are planted in our babies, and they learn that it's OK to drink. For adolescent males, this constant viewing, combined with the nightly ritual of street corner drinking in the hood, serves as a rite of passage.

40oz Rite of Passage

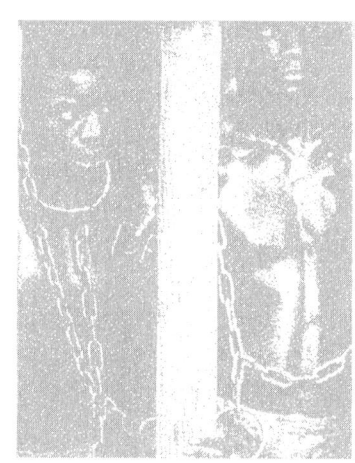

The 40oz rite of passage has become as deadly as the very first passage the olde English provided for us. It was called the middle passage, and it was the ultimate trip. A dead man in a coffin was given more room than the slaves who were chained to one another in the bottom of the olde English slave ships. Enslaved Africans had only a couple of inches to move their heads. As children take up little space, ships built to accommodate 500 were able to squeeze in 800 or more. To the olde English, African bodies simply meant profits, and the more the merrier. For the Africans, however, more bodies meant serious health problems, including yellow fever, malaria, leprosy, measles, smallpox, chills, and infections of the bowels, lungs, and throat–not to mention a spiritual and emotional sickness from which we have not yet healed. The death rate per ship ranged from 15-30 per 100.[42] For eight to ten weeks, our ancestors suffered the unbearable stench of feces, disease, and the horrifying screams of women birthing live and stillborn babies. This hell was justified by the "moral" and religious beliefs of the olde English. The unholy trade continued from the mid 1500s to the end of slavery.

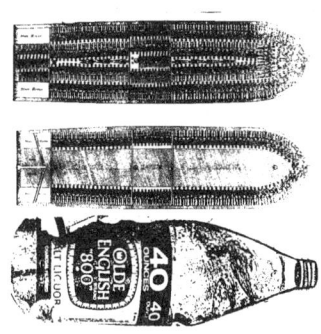

Olde English slave ships–yesterday and today.

Yet, Pabst Brewing Co. honors the olde English by naming their power brew after them. Olde English 800 enslaves our people just as the olde English did, causing illness, pain, and death. Clearly, the sale and promotion of Olde English 800 can be interpreted as a reenactment of the enslavement of Black people. Consider the barbaric legacy of the olde English and their historic interactions with people of color, especially Black people. The olde English used alcohol to enslave, colonize, and kill, then and now.

According to Eric Williams in *Capitalism and Slavery*, "The eighteenth century in England was notorious for its alcoholism."[43] While we often call alcohol *spirits*, the olde English called their brews *kill-devil*.[44] Olde English 800 is a killer spirit, a devil. The devil seeks to kill and destroy. He comes like a thief in the night. Most homicides take place in the night. Most arguments between people who know each other and that erupt into violence began with a couple of drinks. This beast brew knows no boundaries. Poor and rich alike can be taken in by its lure of power and sexual conquest.

Rodney King himself had built up a tolerance for "eightballs" the street name for Olde English 800,

MESSAGE 'N A BOTTLE

> *"Health officials recommend that a person consume 64 ounces of water a day, not 64 ounces of malt liquor. Yet you're not to assume that the malt liquor industry is trying to kill you. Think again!"*
> –Pam Thomas, Dayton, OH

a high alcohol beer. That evening King had consumed enough eightballs–roughly the equivalent of a case of regular 12-ounce beers–to put his blood alcohol level at twice the legal limit. But he wanted still more: when he spotted the highway patrol car behind him around 12:40 a.m. He had just exited the freeway on his way to another liquor store.[45]

And then those police officers beat Rodney King like the olde English used to beat Black slaves. *Source* magazine provides another glimpse into 40oz addiction. This time the victim is a wealthy rap artist.

> DMC's unpredictable nature–coupled with his love for "Olde English 800"–has caused more than just a few problems for the group over the years. For DMC, being a celebrity was cool, but it couldn't compare to the thrill he got when he twisted the cap off of a fresh 40 ounce bottle of his favorite malt liquor. He was rarely seen without a bottle in his hand. He just drank his forty and chilled. He eventually drank his way into the hospital. Suffering from alcoholic pancreatis, DMC found himself faced with two choices: Life or Death. When asked how many forty's did he drink per day, DMC said 12, 13, 14, 15, 16 make it 16.[46]

> *"Olde English 800 was one of the first malt liquors to be marketed to the Black community in a 64oz container. I guess 40oz wasn't killing us (Black people) fast enough. Have you ever seen John Wayne in his tough guy roles ever order 40 or 64oz of whiskey? I don't think so!"*
> –Ronald Kwesi Harris, Chicago, IL

16 x 40oz = 640 ounces of Olde English 800. That's close to 14 six-packs of regular beer!

The Black community's been played big time. Apparently, the alcohol industry understands that as a whole, we are disconnected from the truth about the African slave trade. Many of us don't want to remember. Too much shame and pain (maafa). In *Historical and Cultural Atlas of African Americans,* Molefi Asante and Mark Mattson state that

> England came to dominate the trade by aggressive attacks on other slave powers and intense competi-

64

tion for new regions. So successful were Sir John Hawkins, the originator of the English slave trade in 1562, the English barons and imperial merchants had written the most daring story in the history of European slave trading. As John Hope Franklin wrote in his book prosecuting the slave traffic was the product of English ingenuity.[47]

We know that alcohol is called spirits. Brothers & Sisters, with 6% alcohol per volume in each bottle of Olde English 800, just how big and powerful is this spirit? Is it powerful enough to possess you mind, body, and soul?

Think about it, every time you drink OE, you're consuming the spirits of the olde English: Sir John Hawkings, Henry(s) VI and VIII, Sir Walter Raleigh, John Smith, Cecil Rhodes, King James I, Queen Elizabeth, Queen Victoria, Sir Francis Drake, Henry Hudson, William Bradford, Bartholomew Gosnold, Charles II, George Washington, John and Sebastian Cabot, Isaac Royall, and every other George, Tom, Dick, and Dirty Harry who enforced the laws of the British crown. The murderous spirit of the olde English lives on every time you drink their brew. The olde English committed savage acts under the guise of civilizing and Christianizing the Western world, while preaching freedom of religion. The olde English claimed every inch of land and every human being they could put their foot on (literally) in the Americas and Africa.

> I contend, said Cecil Rhodes , that we are the first race in the world, and that the more of the world we inhabit the better it is for the human race...
> If there be a God, I think that what he would like me to do is to paint as much of the map of Africa British red as possible.[48]

Ironically, a dry blood red color figures prominently on the cap and label of Olde English 800.

Olde English History 101

The olde English comprised a nation of white Europeans who lived on the isle of Britain. This country is also known as Great Britain, United Kingdom, British Commonwealth, British Crown, British Colonies, British Empire, and New England (America).

The olde English ruled the Western world from the late 1500s to the early 1900s. They gave birth to one holocaust after another. Those who survived their savage invasions were forced to submit to the British Crown. Whenever we hear the term "olde English," we should automatically think "historical trauma for peoples of color." Every time we look at an Olde English bottle we should remember slavery, ethnic cleansing (genocide), and colonialism. We should remember war and misery. Instead, we try to forget by drinking a mind numbing brew that was named for the same people responsible for our oppression and enslavement. The Sleeping Giant slumbers on.

The olde English were determined to establish and control a New World Order "for God and the Empire" by any means necessary. The British Empire exceeded <u>800</u>,000 square miles both by land and sea.[49] Question: Why would a company name an alcoholic drink after a nation whose official policy was to terrorize, murder, and steal? Could it be that the spirit and tradition of the olde English live on inside the bottle and Pabst Brewing Co.?

The Crown

On the label, the name Olde English is surrounded by crowns. The combination of the crowns and the Olde English name suggests the British Crown. The olde English bought and sold African slaves with a currency they called a "crown"; one crown equalled 25 pence. The Olde English 800 label bears 33 crowns, 32 of which appear in a circle. To understand the relationship between the crowns and the number 800, multiply the 32 crowns by 25: 32 x 25 = 800. Coincidence?

According to historian Basil Davidson, Britain worked a deal with the Spanish king that gave them a virtual monopoly of the slave trade. "The British were to provide 144,000 slaves over thirty years or an average of 4,800 a year. This monopoly they purchased from the Spanish king for 200,000 crowns and agreed to pay a duty of 33 1/3 crowns for each slave landed alive."[50] Davidson lists the price of a slave at 33 1/3 crowns. W.E.B. DuBois gave the price at 33 1/2, while other historians say 33.[51]

Thirty-three crowns are depicted on the OE label. Thirty-two crowns are strategically placed in a 360 degree circle. Now a crown is often used to symbolize the sun or sun rays in art. Thirty-two crowns arranged in a 360 degree circle symbolizes the sun. And as we all learned in our Eurocentric history classes, the sun never set on the British empire. No continent escaped the domination of British colonialism. No Black community in America has escaped the invasion of Olde English 800.

The Number 800 (and 8)

We learned in chapter 2 that the Greek letter omega (Ω) carries a numeric value of 800. Omega means the "last" or "end." In addition to its traditional meanings, "crown" also meant "funeral" to the olde English.[52] A funeral certainly symbolizes the end of life, which is consistent with the 800 omega symbol. Could it mean that one glass of Olde English 800 malt liquor could be your last?

The descendants of the olde English were the New Englanders, or New Americans. During the American Revolution (1775-1783), the New Englanders often bartered for slaves from the west coast of Africa. The going rate for 40 healthy male slaves was 800 gallons of rum. One healthy male slave cost $800.[53]

The 8-Ball Alias

The name 8-Ball is just one of many aliases used to promote the sale of Olde English 800. It's old marketing tricknology to marry street slang with logos and slogans. In the hood, an 8-Ball is an eighth of an ounce of cocaine. Furthermore, 8-Ball is the name of a very expensive and popular designer leather jacket that has status within the street drug culture.

Webster defines eightball as being "in a tough or dangerous position." When you drink 8-Ball, you put yourself in a tough and dangerous position. The slogan "OE's the name, 8-Ball's the game," plays to the hardcore street mentality.

The slogan has a deeper message for the Black community. In the *Isis Papers*, Frances Cress Welsing describes in great detail the underlying racial symbols of the olde English game of billiards or, as we say in the hood, pool. The movie *Boomerang* helped make her breakthrough interpretation famous. Martin Lawrence's character talks about how racist the game is, with the white ball always dominant and the black ball always in an underdog position. Even Eddie Murphy's fly character has to admit that he might have a point.

8-Ball is often played by brothers in the hood while drinking Olde English 800 and other malt liquors. The object of

OLDE ENGLISH 800

the game appears to be simple. The first player to knock off all the solid or striped balls first, then the 8 ball without the white cue ball falling into one of the six pockets, wins. That's the outward appearance of the game.

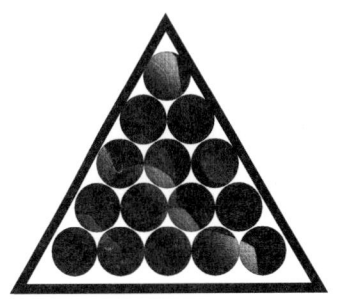

Psychologists often study how children play in order to determine why they're acting out. To understand the abnormal behavior of Europeans involved in the murder and genocide of peoples of color around the world, Dr. Welsing looked at their play, i.e., their many ball games, including the game of pool.

The pool table is flat and green and symbolizes the earth. (Keep in mind that Europeans believed that the earth was flat.) The cue ball is white and powerful; it rules the table (read: earth). The white cue ball represents the white male (olde Englishman). According to the *New Dictionary of American Slang*, "A cue ball symbolizes an eccentric person, an odd ball."[54] The only ball without color (melanin), the white cue ball is the odd ball on the table. It is isolated and privileged during the game (of life). The white cue ball (olde Englishman) has only one job, and that is to rule the table (earth). The white cue ball (olde Englishman) roams the table (earth) with the aid of a superior weapon (cue stick), pursuing, stalking, hunting, and chasing the other balls of color (people of earth) into the pockets. The white cue ball (olde Englishman) seeks to control and dominate with the aid of his superior weapon. The white cue ball is the only ball allowed to use a superior weapon. No other ball of color has this privilege on the table.

At the start of the game, all the balls of color are placed inside a triangular rack (pyramid), which connects them spiritually, for an upward pyramid symbolizes the Holy Trinity. One could say the balls (people of color) are surrounded by the Father, Son, and Holy Spirit. Any and all who reside outside of the rack (the white cue ball, olde Englishman, the alcohol industry, etc.) are antispirit, antilife, and antisocial.

Placed in the center of the balls of color inside the rack (pyramid) is the black 8 ball (Black parent/Black male). The Black male is the only individual on the planet with the genetic power to produce all the colors (races). Thus, placing the black 8 ball in the center of the pyramid (life) makes sense. Furthermore, it signifies that the seed (or womb) from which life comes is black!

Clearly, the black 8 ball can be viewed as a symbolic parental figure from which all life emerges; the white cue ball is the symbolic emotionally disjointed stepchild who is full of rage. The unprovoked white cue ball starts the game by rushing the symbol of life (pyramid), violently disrupting the peaceful family gathering, causing the family to scatter (run for their lives) about the table (earth).

The white cue ball "shoots" each ball of color. It appears to be preoccupied with segregating the table. Once shot by the white cue ball, the balls of color scatter to any one of the six pockets (liquor stores, crack houses, ghettoes, prisons, graveyards, and mental institutions) on the table. Once the white cue ball has imposed its power over the balls of color, it then seeks the title of landlord by assassinating the black 8 ball (original Black parent). In an effort to capture the table and win the game (of life), the white cue ball (olde Englishman) assumes the privilege and power of resurrection.* The white cue ball is the only ball on the table that can be pulled out of a pocket. It assumes this advantage because

---

*Much occult and new age thought deals with the subject of immortality and ascending to the heavenly planes to live with a group of masters called the "White Brotherhood."

there *must* be a final showdown with the black 8 ball.

This unusual privilege was an attempt, conscious or unconscious, by the olde English designers of the game to reverse the symbolic meaning of 8. According to *Number In Scripture*, the number 8 (or 800) is associated with "resurrection" and "regeneration" and the "beginning of a new era or order."[55]

Brothers & Sisters, after all these years of playing 8-Ball we were lead to believe that the game was over every time the black 8 ball (Black parent) prematurely fell off the table (earth). The truth is, the white cue ball (olde Englishman) stole the power and privilege of resurrection and regeneration from the black 8 ball (Black parent), and the world has been fooled ever since. An imposter, the white cue ball must assassinate the black 8 ball so that it can maintain its dominant status on the table (earth). If the black 8 ball stays on the table, the final confrontation for ruler of the table will be black vs. white. If the 8 ball is forced off the table prematurely, the white cue ball immediately terminates the game and declares itself the winner (landlord). The white cue ball's victory symbolizes the goal of *white supremacy**, i.e., the symbolic death of the black 8 ball and the symbolic impotency of the other balls of color. The next time you play pool, watch how shots against the black 8 ball are made with a higher level of aggression than shots made against the other

---

*It is our belief that not all white people are white supremacists. Even white people suffer under the world system of white supremacy. It is a system that will sacrifice its own to maintain global domination and deal with the threat of genetic annihilation by the Black man. However, we also believe that all whites, even well meaning ones, can slip into olde English beliefs (disguised as liberalism), attitudes, and behaviors if they have not done serious study and self reflection. Therapy and consciousness raising workshops, such as Brown Eye/Blue Eye, are highly recommended.

balls of color. The game of 8-Ball symbolizes the olde English (European) desire to rule the earth and the people on it. It also symbolizes the white man's fear of the Black man's genetic ability to annihilate him.

Let's rewrite Pabst's slogan, "OE's the name, 8-Ball's the game," to reflect the truth: "White supremacy's the name, killing people of color is the game."

Since *Message 'N A Bottle* is about healing, we offer the following new game plan for pool players struggling under a system of white supremacy. The underlying theme of our new game is truth and correctness. The new game is called N/A.*

First, I highly recommend that you shoot with the ball that best represents your racial group. If you are African American (Black), shoot with the 8 ball; Native American, the red ball; Hispanic, the brown ball; Asian, the yellow ball. If both players are of the same racial group, flip a coin to see who will shoot first, then share the ball throughout the game.

The mission of the game is to undo the self hate that the olde English game of billiards has reinforced among the races of the world. You will be ridding the table (earth) of warism, racism, sexism, and poorism. The object of the game is to put the white cue ball (olde Englishman) into the pocket of recovery and reeducation, i.e., rehab.

Start the game by placing the white cue ball inside the rack at the tip of the pyramid. But wait a minute: instead of turning the pyramid upward, this time turn it downward, which symbolizes the truth of how white supremacist beliefs and behaviors have dragged down peoples of color into disease, self-hatred, violence, and drug and alcohol addiction.

---

*Here's another clue to decoding the secret message hidden within the title of this book.

Next, identify which two of the six pockets will serve as rehab. These are the pockets the white cue ball will be helped into. Combine the two pockets any way you like, just don't change them during the game. Players who select stripes can only combo with stripes; solids combo with solids. (If you are playing alone, you can combo with both stripes and solids.)

Each player gets three rescues, i.e., any ball of color that is forced off the table prematurely (prior to knocking the white cue ball into rehab) can be resurrected (three balls per player per game.) If you are playing alone, you are allowed five rescues during the game. The reason for combo shooting is to establish and reinforce color-on-color alliance. No one group should be burdened with the awesome duty of ridding the table (earth) of racism, warism, sexism, and poorism alone. Furthermore, once the white cue ball is helped into the rehab pocket, no other ball of color is allowed to join it. This is important! We who have been victimized by the white supremacist media, "just us" system, police, government bureaucracy, health system, and alcohol and drug industries are also in need of rehab and reeducation. We run the risk of deeper infection if we continue to deal with the people (and their products, e.g., alcohol, cigarettes, junk food, etc.) who want us sick, crazy, and dead.

You'll be amazed to discover that putting the white cue ball (olde Englishman) into one of the rehab pockets via combo shooting won't be easy. In fact, the odds of the white cue ball remaining on the table are quite good. It will take a while to master this game, but it will be well worth the time invested. You'll discover just how complex the issues of racism, sexism, warism, and poorism are within a society that is run by white supremacists, and how difficult it will be to rehabilitate and reeducate the white supremacists.

If you want to see how seriously racism and other isms have been internalized by people, attempt to play this game with or around your white friends. Before long, someone will start screaming "reverse racism!" Your friends of color may also feel uncomfortable playing this game.

The Tiger

All the leading malt liquors have a mascot–Colt 45 has a red horse, Schlitz, blue-Black and red bulls, etc. In sports, a mascot brings good luck. A mascot communicates power, terror, and superiority to the subconscious of an opponent.

Pabst Brewing Co. chose a tiger for their mascot. The largest and most powerful of the cat family, the tiger is considered to be unbeatable and is often called a man eater. On middle school, high school, and college campuses across America, the tiger is second only to the eagle in popularity as a mascot.

The Pabst tiger symbol works on many levels. In his book *Culture Bandits*, Del Jones says that "Every communicated message or image has a motive, a target, a perspective. They are created by people who come from a culture, possess an ideology and are discriminating. Therefore, you can throw 'objectivity' out of the window."[56] Consider the following definitions of tiger:

"A strong, virile man; A dangerous man."[57]

"Symbol of ferocity."[58]

"A person regarded as aggressive, audacious or fierce."[59]

"Symbol of wrath and cruelty."[60]

"A person or sometimes an animal of fierce and bloodthirsty ways. A person vigorously aggressive, A groom."[61]

"A fiercely aggressive person in a position of authority. . .a groom."[62]

Keep in mind, Brothers & Sisters, that Olde English 800 is marketed to Black people, with OE billboards strategically placed in our neighborhoods. Why would Pabst use a tiger to market malt liquor to Black people? What are they trying to tell us?

It's tempting to say that the tiger stereotypically represents the Black man as a dangerous, angry animal, and maybe on one level the tiger was meant to work that way symbolically. However, we propose that the tiger was meant to depict the inner nature of the olde English themselves. Why a tiger? Why not a goat or, better yet, a dog? After all, it was the olde English who coined the phrase "A dog is man's best friend."

When the olde English invaded the Indo-Asian region, they disrupted the natural habitat of the tiger when they cleared out the forests (for capitalistic reasons). They then made a sport of hunting the tiger and killing him. The olde English have a history of killing humans and animals for sport. This appears to be some sort of white supremacy ritual.

Among the native peoples of the region, the tiger was both revered and feared. So when the olde English defeated the tiger community, they became the big bad cat to the people. This was a deep form of psychological warfare, which only made a bad case of ego disorder worse. The olde English became obsessed with the sexual prowess of the male tiger. They sold the tiger's penis in a self-created underground market as an aphrodisiac. Once again, the sick need of the olde Englishmen to castrate other males, regardless of the species, rose to the surface. Freud's theory of penis envy does not apply to women; it applies to white men!

Sex subliminals are constantly used in alcohol ads. Even the bottles are phallic in shape and the liquid contained inside resembles either semen (clear, white) or urine (golden yellow). Alcohol companies spend a lot of money to convince you that their products are aphrodisiacs. Nothing could be further from the truth. Oftentimes the male can't perform sexually after he's had a couple of drinks. We're talking chemical castration and *impotence*, which is the exact opposite of "making your jimmy thicker." I believe that the sale of those huge 40 and 64oz bottles of high octane malt liquor is yet another attempt to castrate males of color. And while males of color have been rendered chemically impotent, females of color are riding on the back of the tiger mascot (white male/olde Englishman) on ghetto billboard ads across the country. What's up with that?

If you will recall, the tiger was defined as a "strong virile man" within the olde English culture. Tiger also means "groom." Webster defines a groom as a "young male, an adult male: man, fellow: a manservant: A bridegroom." So there are two possibilities: the olde Englishman as manservant or bridegroom. Now think, have you ever known a white male to be a servant to a woman of color? That leaves us with the tiger (white male) as bridegroom. In the ads, the Black woman is dressed in a sexy string bikini, straddled intimately across the back of the tiger (white male). She is the "bride." This is straight up jungle fever! Decoded this ad reads, "The tiger (white male) gives you, Black man, alcohol for your pleasure while taking your Black woman for his." Here are some more thoughts:

- If a tiger is fierce and cruel, can a woman really safely ride the back of such a dangerous creature?
- A tiger is also called a pussy cat. One of many degrading slang words given to women is "pussy." **A woman (pussy) + a tiger (pussy) + the slogan "It is the power" = "pussy" power.** On a symbolic level, alcohol has the same power as sex ("pussy"). If you become addicted, you're hooked.

Brothers & Sisters, symbols are powerful. You will never see a blonde-haired/blue-eyed woman riding the back of a black panther to promote alcohol.

The Slogan

Pabst Brewing Co. was forced to drop the slogan "It's the power" because federal regulations state that alcohol advertising should not imply strength. It's true, however, that Olde English 800 malt liquor packs a serious punch. At one time it was the strongest malt liquor on the market, doping up at 6.0% alcohol per volume. Regular beer runs about 4.6% by volume. But what if Pabst really wasn't pushing the strength of its malt liquor? What if the slogan was meant to convey something else? I believe that the "It's the power" slogan refers to the power of the olde English and much of the world's submission to the British Crown–not necessarily the power of the brew.

Let's work the slogan. Combine the two words ("it's" is the contraction for "it is"): it is = itis. Next, add to the beginning of that group of letters the first two letters of the word "Britain." Br + itis = Britis. The slang for Olde English 800 is 8-Ball. Place the alphabet equivalent of the number 8, which is "h," next to the letter "s." Britis + h = British!

Message 'N A Bottle

Perhaps the slogan "It is the power" really means "Britain, the power!"

Pabst changed the slogan to "It is the tiger." My Brothers & Sisters, for the love of our community, family, self, and children, let this tiger go!

Consider this rap-poem the next time someone offers you a 40 of Olde English 800 or the next time you see one of the many OE billboards.

OE, Not For Me

Drink "Olde E"...nigger please!
That's the slave masters' cup of tea
If you knew your Black history
The ways of the Olde English, G
The building of the 13 colonies
From the 15th to the 18th centuries

The English made 800 trips to the Motherland, G
John Hawkins and his crew–traded guns and brew
To the slave catchers crew
To catch Black folks like me and you
The English stole our people in the midnight hour
Now they brag "It's the power"
800 means they have come to collect
The plan for slavery is back in effect

800 gallons of rum was the going rate
The price they paid, for 40 male slaves
And you wonder why its called a 40 ounce
Hell! You figure it out

8-Ball means bad luck!
The Black woman was sold for 800 bucks

OLDE ENGLISH 800

The Black man and Red man bow to the English crown
If they didn't, they got beat down!

The English stole our people in the midnight hour
Now they brag "It's the power"
800 means they have come to collect
The plan for slavery is back in effect.
© 1995 Alfred Powell

1854404

WANTED

NAME: Red Dog

OCCUPATION: Assassin

MISSION: To cause spiritual depravation, emotionalism, sexism, sensationalism, denialism, egotism, racism, poorism, and alcoholism.

ALIASES: Bad Dog, Big Dog, Top Dog, Ruff Neck, Bull Dog

DISTINGUISHING MARKS: Mug shot of a bull dog, subliminally pornographic logo

VITAL STATISTICS: ht-12", wt-12, 22, 40oz; 4.6% alcohol by volume. Cheap: $1.35.

CRIMES: Pornography, disrespects women

AFFILIATIONS: Miller Brewing Co., a.k.a. Plank Road Brewery

M.O.: Suspect claims to be "uncommonly smooth." Sexist attitude appeals to men.

LAST SEEN: Suspect last spotted running wild on college campuses. Eye witnesses report seeing suspect on billboards, TV commercials, tee-shirts, and ball caps.

CAUTION: Call the pound, this dog needs to be on a leash. Better yet, put him to sleep!

4. That Low Down Dirty Dog

I know Red Dog isn't a malt liquor, but because of its popularity in urban America and among urban teens I've got to tell you about the covert and overt themes I've discovered that are used to promote and sell this dog.

Drive down any major highway in America and you're bound to see a red dog on the loose. He's on billboards everywhere, telling us things like "you are your own dog" and "be your own dog." As for TV, he's marked territory on NBA and NFL games. This dog's got attitude and a strong, cocky masculine presence. On TV commercials, veteran actor Tommy Lee Jones speaks for the dog in a bold and gravelly voice.

The dog is red. The color red suggests violence, anger, hostility, and aggression. And of course, according to the olde English, a dog is man's best friend. Clearly, the dog mascot is geared to a male market and caters to the cocky attitude of the male ego.

Red Dog beer and the red bulldog mascot have a strong appeal to youth in urban America. They call it Ruff Neck on the street. In fact, several of the slogans for Red Dog refers to the dog as being rough. You'd better believe, Red Dog ain't no Spuds MacKenzie (another alcohol character designed

Hmmm...
Is it a coincidence that Miller launched Red Dog in 1994, the Chinese Year of the Dog?

to appeal to youth). He's just a new dog performing an old trick. This dog symbol appeals to men of all cultures. For example, young Black men in urban America have adopted the slang "dogg," as in "My dogg, what's up?" or "What up, dogg?" (translation: "What's going on?"). Other expressions are "That's dogg" ("That's bad," "That's messed up") or "He dogged me out" ("He mistreated me.").

The olde English may have believed that dogs were their best friends, but today, the dog is the universal symbol for sexually unfaithful men, which is the exact opposite of dog's traditional symbolism of loyalty and faithfulness. Men are called dogs and women are called bitches (female dogs), but are we each other's best friend? Exactly the opposite. Back in the day, if a brother called me dogg, he was looking for a fight. Today, it ain't no thang. It reminds me of how some Black folks like to call each other "nigger." The white slave owners called the enslaved Africans niggers, so we so-called free African Americans call each other by the same name that caused so much fear, hatred, and shame among us. "What up, dogg/nigger?" "You *my* nigger/dogg." What's wrong with us? Have we completely lost our minds?

Consider the following dog phrases and their association to the human male:

Dog Phrase	**Human Male**
hound dog	male who hunts females
dirty dog	unfaithful male
lazy dog	chronically unemployed male
no good dog	low down, worthless male
mean dog	violent, homicidal male
filthy dog	foul, unclean male
crazy dog	verbally and physically abusive male
sly dog	player, pimp, mack daddy
atomic dog	a great male lover
junk yard dog	rough neck male

RED DOG

The dogg mentality is very popular right now in urban culture. Hip hop artists, a.k.a. gangster rappers, are obsessed with this dogg theme. Their lyrics are usually about the doggy style sex act, which is often explicitly acted out in their music videos by the males and multiracial casts of half naked females. Don't be surprised if some brewer produces a malt liquor called Atomic Dog. Just remember, you heard it here first.

While working on an alcohol prevention campaign called "Catch the Attitude", my partner Shack surveyed 100 Black males, 14 to 19 years old. He asked them to state why they were so attracted to the red bulldog mascot. He concluded that the young brothers "overwhelmingly identify with Red Dog's defiant mentality and the appeal of being your own boss." But as brother Shack says, the only thing this dog will fetch for you is the disease of alcoholism.

Brothers & Sisters, Red Dog will never be your best friend. He will never roll over and play dead for you, but he can cause you to roll over and *be* dead. Think about it.

The "Bitch" Commercial

In the summer of 1994, a TV commercial for the promotion and sale of Red Dog beer aired. The commercial subliminally asked the question, "Is the Black woman a bitch?" Yes it did!

As the commercial unfolds, we see the world through the eyes of the defiant red bulldog mascot, who is sitting casually on the front porch of a house located in what appears to be a suburb somewhere in America.

Several women in short shorts ride their bikes past a white picket fence; they are being chased by a pack of dogs. As the commercial progresses, three males, including a white mailman (the symbolic bearer of bad news) and two white men dressed in dark suits and carrying black brief cases (authority symbols), are shown trying to enter the front yard. All three males are hastily chased out of the front yard by the threatening barks of Red Dog. Red Dog successfully defies authority, conveying the impression that he is the top dog in charge.

At the end of the commercial, a female bulldog (we know she is female because she wears a pink bow around her neck) is shown entering into the front yard. She approaches the red dog with the greeting, "What's happening, baby?" This female dog sounds suspiciously like a Black woman. Who else on the planet, apart from some wannabes, say "What's happening, baby" in that unique, sensual way? And who, nine times out of ten, does she say this to? The Black man! Keep in mind that all the roles up to this point in the commercial have been clearly defined as white. White people have been portrayed as sex symbols, messengers, and authority figures. The voices that *sound* Black have been assigned to the dogs, male and female. Female dogs are called bitches, and so are Black women–often by some Black males.

In another TV commercial for Red Dog Beer, the voice of Red Dog states, "I don't go chasing after anything that doesn't have soft hair and big eyes." Then the same sexy Black woman voice states, "Come here, daddy." The M.O. in these commercials is the same. Males are authority figures playing fetch with their dogs. Only the Black woman is depicted as a "bitch."

Once, twice, three times a (bitch) lady? Will there be a third Red Dog commercial? Will the bulldog with the pink bow actually be called a bitch?

The mother of civilization, a bulldog, a bitch.

The Red Dog Sex Scandal

Miller Brewing Co. played us big time in its urban Red Dog billboard campaign. To introduce the product, for weeks, all we saw was this red bulldog–no words, no clues, no nothing. We knew that a lot of money had been put up, but we didn't know who was fronting the bill or what the dog was all about. Was it a movie? Another brand of gym shoe? No one knew, and Miller succeeded in phase one of its campaign to grab the attention of ghetto youth (and adults).

Later, we learned that this was yet another alcoholic beverage being sold to us, but by then it was too late. The damage had been done. Everybody was into Red Dog. What was the appeal of this dog? What kind of spell had it cast on the Black community?

Then, on March 23, 1995, the *Chicago Sun-Times* ran a very interesting piece by Richard Roeper that shed some light on the mystery. Roeper was in a bar one day when, encouraged by the bartender, decided to try some Red Dog beer. Says Roeper,

> There's no way to put this delicately so let me just preface it by saying that the following analysis of the half covered, turned upside down red dog bulldog may not be suitable for younger or more sensitive viewers. What I saw was the head of a nearly bald man between a pair of legs. In the midst of an intimate act.[63]

He then showed the image to other bar patrons; most of them saw the sex act.

It is not my intention to be vulgar but to let you know how you're being manipulated by this dirty red dog. We often refuse to consider new ideas from other Black people, but if the media reports a thing, it becomes believable. Many of us are still "thinking" out of the negropean portion of our brains. Maybe now we'll believe there's more going on in these ad campaigns than meets the conscious eye.

In the decoding business, we look at images everywhichway: upside down, inside out, mirrored, side to side. In light of the oral sex subliminal, the slogans "uncommonly smooth," "slides down real easy," and "the dog comes ruff," make sense. The million dollar question is, Was this image deliberately designed, or was it just a fluke?

Back to the commercial. Remember, the white females riding past the white picket fence (an innocence symbol) while being chased by a pack of dogs (men)? This short but powerful scene was orchestrated to subliminally suggest a group of men (dogs) chasing innocent women ("pussy"). Once again the sex theme has cleverly found its way into your mind and living room. Think my mind is in the gutter? Actually, Plank Road Brewery, a.k.a. Miller Brewing Co., is banking on *your* mind being dirty–to the tune of $15 to 20 million that they have invested into advertising and marketing.[64]

When I first discovered this subliminal X-rated peep show in late 1994, many people thought I was crazy. They said "People don't think like that." Some said I had too much time on my hands. It helped to have the support of a mainstream columnist. I had no problem believing that Miller was capable of selling pornographic beer. My experiences decoding Zima, ST Ides, Olde English, and others taught me that these alcohol companies will sink pretty low to capture market share. It's corporate warfare, and Black youth are the pawns.

Question: Who do the legs belong to, a male or female? The confusion of sexual identity suggests bisexuality. This would be profoundly consistent with the covert attitude of the Red Dog marketing scheme, which is to be your own person, and to break all the rules of the majority.

And then one day I saw it. It had been staring me in the face all along. Turn the bottle upside down and place your index finger across the eyes of the dog. The words "Red" and "Dog" are now upside down. Backwards, Red Dog spells GOD DER.

R E D D O G (forward)
G O D D E R (backward)

Say the backwards version aloud a few times real fast. "GOD DER" sounds a lot like "got her," which is slang used by males to convey sexual conquest. The subliminal legs belong to a woman. If nothing else, the label suggests that it's a dog-eat-dog world.

Beware of Dog

My 17-year-old daughter, Natasha L. Powell, offers these words of wisdom to her peers who may have Red Dog's leash around their necks:

> He is one bad dog, Red Dog that is. He answers to the names Red Dog and Ruff Neck. He runs rampant through institutions of higher education. He has become top dog at campus parties. He has moved from the dog house to the frat house, from the dog pound to the school grounds.
>
> Red Dog is an attack dog turned loose on Generation X and the members of the Hip Hop Nation. This dog has become man's best friend in the worst way. Beware, this dog can't rescue you from the hell he can lead you to.

19-20-945-19

W A N T E D

NAME: ST Ides

OCCUPATION: Tempter, Spiritual Assassin

ALIASES: Crooked I, Jimmy Juice, Snake Bite, Satan Ides, Angel of Death, and Funky Lightning Bolt

DISTINGUISHING MARKS: Two palm trees, a lightning bolt, and an upside down pyramid

VITAL STATISTICS: ht-12", wt-40oz; 8% alcohol by volume. Cheap: $1.39.

NATIONALITY: European

CRIMES: Suspect disguises itself as a saint.

AFFILIATIONS: G. Heileman Brewing Company, McKenzie River Corporation

M.O.: Suspect claims to make your "jimmy" (penis) thicker. Promises to blow you away.

LAST SEEN: Suspect was spotted in the hands of inner city youth and high school dropouts.

CAUTION: Suspect loves to party with "niggas" and wannabe gangbangers. Considered to be armed and dangerous.

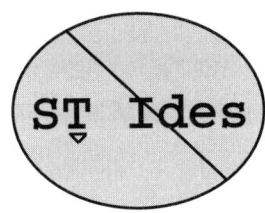

5. Satan'N A Bottle

IDES:
I = I
D = destroy
E = every
S = soul   or   *"I destroy every soul."*
                      –Satan, the devil

Black community activists and grassroot organizations across America have been raising hell for years trying to combat the invasion of advertisers who litter our neighborhoods with their mind control messages of gangs, sex, and violence on billboards and TV. For the most part, their efforts have fallen on deaf ears, mainly due to the powerful financial influence the alcohol industry wields with politicians, our so-called leaders, and the media.

All hell broke loose, however, when a famous rapper named Ice Cube released his album *Death Certificate* in 1991. Labeled antisemitic and antiAsian (Korean), the album was widely condemned by mainstream America. (Interesting that no one ever condemned it for being antiAfrican American.) At least we know now that someone is monitoring rap, which stands for the (R)=revolution of (A)=African (P)=people.

Shortly after the release of *Death Certificate*, the Korean and Jewish communities unleashed a national boycott against products endorsed by Ice Cube, including ST Ides malt liquor. The Korean American Grocers Association (KAGRO) led an estimated 3,000 stores in the boycott. In a show of unity and protest, Korean merchants pulled Ice Cube's poster from their storefront windows. Reportedly, many of the Korean merchants refused to accept new deliveries of ST Ides malt liquor and returned caseloads of unsold ST Ides to their regional distributors.[65]

Needless to say, the protest caught the attention of the McKenzie River Corporation. ST Ides is a registered business of G. Heileman Brewing Company. It is brewed by Heileman's Blitz brands and marketed by the McKenzie River Corporation. McKenzie River was launched in 1987 as a business venture between former Blitz-Weinhard Brewery owner Minott Wessinger and San Francisco-based public relations firm Goodby, Berlin and Silverstein (GB&S). Together they created a highly successful national advertising campaign targeted to African American males.[66]

McKenzie River answered the concerns of the Korean and Jewish communities by agreeing to put Ice Cube on ice temporarily. Furthermore, the company pledged to pay off KAGRO $90,000 to end their boycott.[67] McKenzie River didn't offer the African American community a dime when grassroot activists complained about the marketing of their upstrength brew (8% per volume), but they acted with lightning speed to pacify the Korean community.

The Campaign

Few people suspect that many of the ads in the ST Ides malt liquor marketing campaign are full of satanic themes and symbols.

ST IDES

ST Ides, the name, has an unmistakable religious connotation. Webster says that a saint is "a soul gone to heaven, especially (Roman Catholic and Orthodox churches), a person whose holiness has been attested by miracles after his death and who has been officially recognized (canonized) by the church as worthy of veneration and to receive intercession (abbr. S. or St, pl. SS or Sts)." Given the alcohol industry rule that religious themes never be used in the advertising of beer, why would McKenzie River/G. Heileman christen their alcohol product *ST* Ides? We may never know the company's true intentions (beyond profit), but if the company is being obedient then *ST* cannot possibly stand for saint. But if *ST* doesn't stand for saint, what does it mean?

The words "street," "state," and "stereo" are just a few of the words that might use the *ST* abbreviation, but they don't really make sense in the context of malt liquor. A look through the *s* section in the dictionary would reveal too many *s*'s to list here, so let's cut to the chase. Brothers & Sisters, I propose for your consideration a couple of thought-provoking *ST* words that should make this discussion very interesting: *Santa* and *Satan*. Note: Santa Claus is often referred to as *St.* Nicholas or *Saint* Nick.

Santa and Satan share many of the same characteristics. On a basic level, their names share the same letters. Both are masters of deceit and temptation. Satan has a band of rebel angels that assist him in carrying out his evil schemes; Santa has a band of elves (little helpers who work like the devil) that assist him. They both make promises they can't keep. Satan and Santa are counterfeiters. The spirit of Santa is so powerful that adults lie to children to cover for him. Grown folks pretend to be Santa-like rather than Christ-like. Satan and Santa deliberately compete with the word and works of Jesus Christ. In the meantime, children become mentally distressed, or even evil spirited, upon learning that Santa is a figment of man's imagination. Like Satan, Santa causes severe depression. The "holiday blues" are common around Christmas.

In some countries around the world, Santa Claus is a religious icon. In Holland, the devil is called "Old Nick." Furthermore, the name "Kris," as in Kris (Chris) Kringle (a.k.a., Santa Claus), rivals the name of Christ.

Santa, like Satan, has the ability to change his appearance; Santa appears as a saint but behaves like a devil.

ST transforms a mere malt liquor label to one that is loaded with religious symbols and connotations. Saint never appears spelled out on any label or promotional material (although in many of the radio and TV commercials, *ST* is stated as *saint*). When we see *St*, however, we automatically think *saint*. Our brains are funny that way. We often see things that are not there. We'll fill in the blank and empty spaces.

For argument sake, let's explore the possibility that *ST* stands for the one and only saint of evil, Satan himself. A small downward pyramid appears under the letter "T." This symbol has strong Satanic overtones. According to Liungman, "Xenocrates, who died in 324 B.C., stated that Δ was a symbol for God. In Christian symbolism it stands for the Holy Trinity."[68] If the upright pyramid symbolizes the Godhead of Christianity, the downward pyramid must symbolize God's opposite, the devil and his legion. The downward pyramid reinforces our suspicion that the symbol *ST* stands for Satan. Consider this: the leading themes in alcohol ads and youth culture music are satanism, death, sex, and bisexuality (more on this in vol. 2 of *Message 'N A Bottle*).

Who (or What) Is ST Ides?

Back in 1992 (maybe '93), I called up the public relations department of McKenzie River and asked the following question: "Where did the name *ST Ides* come from?" One of the first theme songs for ST Ides malt liquor was sung by Levi Stubbs of the Four Tops in 1987. The song's tropical flavor strongly suggested a Caribbean connection. I was surprised to hear that the original name choice was not Ides but *Ives*. Unfortunately for McKenzie, the name was already attached to a lotion product. Their second choice was Ides. In other words, according to McKenzie's public relations person, the name was "invented" and "it doesn't mean anything."

Or does it?

Let's play a word game to help us establish a symbolic identity of ST Ides. Combine the two names (the original and the invented): IVESIDES. What words are hidden here?

IVESIDES
I V E S
I D E S
D E V I

Four of the five letters almost spell *devil*, and there's even a symbolic letter *l* to complete the word. Note that *I* in Ides has a distinctive design. It looks like a lightning bolt, and could thus represent either an *I* or *l* (for lightning).

IVESI(L)DES
I V E S
I (L)D E S
D E V I L

MESSAGE 'N A BOTTLE

"Now it makes sense to me why they spell the word Ides with a lightning bolt. I knew they had a reason for it. This workshop has helped me to confirm my suspicions about ST Ides."
–Jose Rodriguez, 15, Port Authur, TX

The word "devil" is hard to miss. Just who is this ST Ides with the crooked *I* or *l*? "That old serpent called the devil, and Satan which deceiveth the whole world" (Revelation 12:9).

Evil lives in the world, and the saint of all evil is Satan the devil. Note that the word evil is contained within the word devil. Targeting the sale of alcohol to a specific group, especially youth, is simply *evil.* Furthermore, evil spelled backwards is live, in other words, *evil lives.*

ST Ides, Counterfeiter of the Holy Spirit?

The Bible says there is one unforgivable sin, and that is the blaspheming of the Holy Spirit. Mark 3:29 (LV) calls it an "eternal sin." Think for a moment: Alcohol is often called a spirit. A saint is a holy person, and an upright pyramid can symbolize the Holy Trinity, a member of which is the Holy Spirit. Thus,

**alcohol/spirit + downward pyramid + ST Ides/devil =
the blaspheming of the Holy Spirit**

Is the McKenzie River Corporation committing the only unforgivable sin, i.e., blasphemy of the Holy Spirit? And what about the brothers and sisters who drink this unholy spirit? Are they also committing an unforgivable act?

In an interview given to the *San Francisco Chronicle*, Minott Wessinger claimed ST Ides is the strongest malt liquor on the block. "That's important," said business partner Andy Berlin of GB&S "because malt liquor drinkers are looking for a beer that gets the job done."[69]

Gets what job done, Brothers & Sisters? Alcohol does a pretty good job of causing illness and death in the Black community. Is that what they're talking about?

ST IDES

When Those Saints Come Marching In

With all that said, what if *ST* really does stand for saint? Brothers & Sisters, what do we call each other in church? Saint! The preacher prays for the "saints of God." The Black community is spiritually rich, and the old church song, "When those saints come marching in," clearly describes the African American spiritual experience. Dick Gregory has long said that people abuse drugs and alcohol because they want to feel and know God, and the high is like a spiritual experience–temporarily. Black people want to be in that number with the saints, so McKenzie River provides us with ST Ides malt liquor to quench our desire to march into heaven. But the truth is, addiction to alcohol is a march straight through hell!

History shows that whenever Europeans invaded a land they would usually march in the name of a saint, Jesus, Mary, or the pope. Their flags would often bear the sign of the cross. The slave ship *Jesus* carried hundreds of Africans to a fate worse than death.

The name ST Ides manipulates us spiritually and emotionally. It evokes the image of godliness, goodness, and holiness, thanks to the *ST* which implies *saint*. Never forget that historically, people of color have been forced to accept the European Christian version of sainthood under threat of death!

Brothers & Sisters, beware this ungodly saint called Ides. It brings death and destruction to our community. To reject this saint would be a high spiritual act of personal salvation. No one should have a spiritual commitment to a "saint" that promotes drunkenness. Even if you are naive enough to believe that ST Ides was named after a real saint, you need to know that his power is pseudo. He can't make you strong or sexually powerful. He can, however, make you sick and dependent. ST Ides pimps you on a psycho-spiritual level.

You'd better ask somebody before you submit, commit, and worship this devil in disguise.

Beware the Ides of March

The Ides of March is associated with *Julius Caesar,* Shakespeare's famous play, and *Life of Julius Caesar*, the biography by the Greek philosopher Plutarch.

In Shakespeare's version of events, Julius Caesar was warned by the soothsayer* to "Beware the ides of March." Caesar ignored the warnings and many other strange signs of his impending murder. He was killed on March 15, the day known as the Ides of March.

Thus *Ides* has an historical link to violence, murder, and death. The owners of ST Ides chose to use this word on an alcohol product that is marketed exclusively to communities of color where violence, murder, and death are common, where money and food are low, and where frustration is high, especially during the middle (the 15th) of the month.

Ides can be decoded to mean "the day of death." If a saint is holy and Ides is the day of death, then we can interpret ST Ides to mean the "holy day (holiday) of death." You don't have to be a rocket scientist to figure this thing out. Or is it all a coincidence?

Let's now decode Ides by working the letters. Ides is an acronym relating to the Black experience with alcoholism:

> I = Intellectual
> D = death (of the)
> E = emotional and
> S = spiritual Self   *or*

---

* Soothsayer: One who interprets dreams.

ST IDES

> *The intellectual death of the emotional and spiritual self.*

Ides can also be reworked to form these words and prefixes:

DESI Alcohol becomes a *desi*re.

DESI Begins "*desi*gn." ST (Satan) Ides has *desi*gned a plan for your destruction.

DIE(s) When you *die*, you stop living.

DIES *Dies solis* is Greek for the day of the sun, or Sunday.[70] Saints are usually worshipped on Sunday.

DISE Begins *dise*ase. Alcoholism is a *dise*ase.

DEIS *Deis*m is a belief in God.

SEDI Ides spelled backwards is **SEDI**, which is the root of the word *sedi*ment, which means "to settle to the bottom," as in life. Addiction to alcohol makes you settle to the bottom of life, i.e., the gutter.

SEDI *Sedi*tion means to incite hostility. Alcohol leads to hostility.

SIDE A saint must choose which *side* he's on, good or evil.

Brothers & Sisters, beware of the Ides before it rolls down on you like it did Julius Caesar and his crew.

Satan: Tempter 'N A Bottle

Have you noticed that the abbreviation for saint within the ST Ides promotional scheme appears capitalized? This is very strange. Saint is usually abbreviated *St*, with a lower case *t*. Furthermore, the abbreviation *ST* features an upside down pyramid under the letter *T*. This upside down pyramid provides us with strong evidence that ST Ides is really Satan Ides. Many scholars will argue that the word Satan is abbreviated as *S*, not *ST*. However, we propose that the *T* might

stand for something else, something related to Satan. Question: What is the weapon that Satan uses to carry out his evil plans? Answer: TEMPTATION![71] Temptation entices you to do evil. Satan never leaves hell (home) without his weapon Temptation. Satan tempted Jesus Christ on the mountain for 40 days and nights (Luke 4:2), and he tempts you with 40oz of malt liquor. Symbolically, Satan appears this way on the ST (Satan) Ides label:

```
S  T
A  E
T  M
A  P
N  T
   A
   T
   I
   O
   N
```

Notice the subtle appearance of the abbreviation ST (saint) when the words Satan and temptation appear side by side. The reason so many of us give into Satan's temptation is because he appears to us as a saint. Brothers & Sisters, ponder 2 Corinthians 11:14, "And no marvel; for Satan himself is transformed into an angel of light." Remember, Satan has the power to appear as a saint (Lucifer), but he's only a devil that hides behind the halo of intoxication!

The Lightning Bolt

The letter *I* in ST (Satan) Ides never appears in standard letter form but always as a fashionable lightning bolt throughout the promotional scheme. In a radio spot for ST (Satan) Ides, Ice Cube and EPMD end their rap with the phrase, "Give us one of those funky lightning bolts," i.e., a container of ST

ST IDES

(Satan) Ides malt liquor.

The lightning bolt has a profound Satanic association. "Lightning from the sky separates into branches; hence, the lightning bolt found in the hand of the supreme god of the sky was stylized into a pitchfork. This was a symbol for his power, as he who controls the lightning bolt."[72]

Many of the print ads and posters for ST (Satan) Ides malt liquor only feature the abbreviation *ST* (in bold print) with a crooked lightning bolt beside it and sometimes below it.

ST ⚡

The puzzle looks like this:

ST + ⚡ = Satan + Lightning

This not-so innocent symbolism is profoundly consistent with scripture. Luke 10:18 says, "And he said unto them, I beheld Satan as *lightning* fall from heaven."

The lightning bolt has a zig zag shape. A zig zag by any other name is still a snake in the grass. "In many cultures lightning is portrayed as a snake cast down from heaven."[73]

Zig zags, or lightning bolts, are also used in comic strips to convey wrath and anger.[74] The bolts are usually drawn close to the head of the characters. This is a subtle but powerful symbolic message. Every time you drink from a container of ST (Satan) Ides malt liquor, the zig zag/lightning bolt is placed close to your head, symbolically indicating that you are angry, violent, hostile, and dangerous. These are the very terms used by the media to depict young Black men. Coincidence?

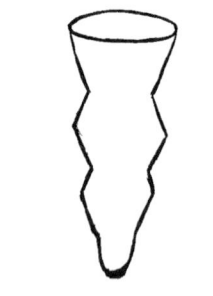
Jimmy hat condom

Jimmy hat cap

Furthermore, this lightning bolt is now on caps, which is worn where? On top of the head! Hello. Many of these "innocent" caps were given away free in some areas around the country. Rap tapes suggesting that real men drink ST (Satan) Ides malt liquor were also given away. We don't know if the caps were bootleg or if the promotions were official, but it does make you wonder: Were young Black men targeted to receive these free promotional giveaways?

To all of you lucky winners of ST (Satan) Ides caps, consider this: In the hood, the zig zag also symbolizes a jimmy hat condom. This symbol on a cap/hat makes it a "jimmy hat"–and you a "dick head." Could it be that you've been played? Brothers & Sisters, "The entrance of thy words giveth light; it giveth understanding unto the simple" (Psalms 119:130).

One final word on the lightning bolt, or crooked *I*. There is a ST (Satan) Ides poster that has become very popular among teens in urban cities. The poster features the word CROOKED written in big bold letters. Just below it is an oversized zig zag symbol in red and trimmed in blue with the background color a solid black. Symbolically this poster connotes a crooked serpent. This poster is profoundly consistent with scripture. Job 26:13 says, "By his spirit he hath garnished the heavens; his hand hath formed the Crooked [swift] serpent." The big bold zig zag or lightning bolt on the ST (Satan) Ides poster is the same symbol used throughout the ages to signify that Satan's job was to make the perfect and upright man Job crooked. Many a brother and sister has been made knocked down falling out crooked in the gutter because of an addiction to ST (Satan) Ides and other malt liquors. Satan in a bottle is doing his job.

The Palm Trees

Have you noticed that two palm trees appear on the label of ST (Satan) Ides malt liquor? The palm trees appear to be bending over from left to right. Why use palm trees in the logo?

Here's what the Bible says about the palm trees: "The righteous shall flourish like the palm tree" (Psalm 92:12). The Bible makes the palm tree a holy symbol when it compares it to the righteous. The palm tree also symbolizes the paradise to come after the last judgment. Palm Sunday is an important holy day in Christianity; it is associated with Christ's triumphant entry into Jerusalem. A palm tree connotes spiritual victory.

If *ST* stands for Satan, and a palm tree (two in fact) symbolizes a spiritual victory, then who is being depicted as victorious on the ST (Satan) Ides label? Satan! And I hate to admit it, but this symbology might have a point. Brothers & Sisters, every time you put a container of malt liquor to your lips, every time you allow addiction to take over your mind, body, and spirit, every time you kill your brother under the influence, every time you have unprotected, promiscuous sex-you lose and Satan wins!

The palm tree is a strong tree. It "will grow tall even if a heavy weight is suspended from it. . .*the palm tree bends not*, staunchly bears the weight."[75] The palm trees on the label of ST (Satan) Ides malt liquor are shown bending from left to right. This symbolically signifies that something is wrong in paradise. Furthermore, "C.G. Jung sees the shape of a palm tree as a symbol of the mind."[76] Thus,

**Bent palm trees + the mind =
a mind blowing experience**

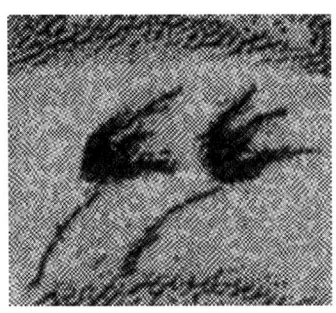

"Notice the label for ST Ides. It looks as if a violent storm is taking place–blowing trees, a powerful lightning bolt, and a black skyline. It appears that 'all hell is breaking loose,' and it does after you drink it."
–Munirah A.L. Bakari, Milwaukee, WI

Brothers & Sisters, the opposite of paradise, or heaven, is hell! The ruler of hell is none other than Satan (the bad *St*). Alcohol, like Satan, is a master of deception. Both are deceivers. They offer you paradise but give you hell! This is no joke. You'd better read the palms of ST (Satan) Ides.

One final point about the palm tree logo. Relax your eyes and look at the palm trees on the label. Focus on the left side of each tree. An outline of a face appears in each one. The strange looking faces (spirits) look as if they are being blown away. You decide how they were blown away, and keep in mind that alcohol was involved. After looking at the hidden faces on the label one understands why McKenzie River chose the Geto Boys hit rap song, "My Mind is Playing Tricks on Me," to promote ST (Satan) Ides malt liquor.

The Jimmy Hat Condom (Say It Ain't So!)

If you think that what you have read so far has been insulting, check the language and symbols McKenzie River has Black youth idol, gangster rapper Ice Cube, singing to promote ST (Satan) Ides!

Thus far we know that the letter *I* has been used as a gimmick throughout the marketing scheme for ST (Satan) Ides. The letter *I* has been decoded to mean lightning bolt and serpent. Now let's find out whether the bold zig zag has any other meaning on a cultural level, specifically urban culture. In the commercial "Ice Cube at the Apollo," the brother raps the following lines:

> *Get your girl in the mood quicker,*
> *and get your jimmy thicker*
> *with ST Ides malt liquor.*

"Get your girl in the mood quicker, and get your jimmy thicker are words designed to kill "niggas"(ignorant people). Every shortie in the hood last summer was rapping these genocidal words."
–Marlon Shackelford, Dayton, OH

Understand that the commercial is not talking about brother Jimmy down the street. Jimmy is street slang for penis. Jimmy is also slang for condom. The lyrics are overtly sexual and totally inappropriate for the intended audience, which, we must assume given the fact that *teen* idol Ice Cube was used, are Black youth.

ST (Satan) Ides malt liquor aims straight for the sexual ego (phallus) of the Black male, this despite federal regulations that prohibit obscene or indecent alcohol advertising. (If that's the case, darn near every alcohol product on the market has broken that rule.)

The next time you see a brother drinking a 40oz of brew, check out his body language. One hand will be gripping the bottle while the other, nine times out of ten, will be on or near his penis.

The joke in the Black community is that brothers are always holding onto their penises because they want to make sure that they're still there. Given our history with lynching in this country, it's a natural response and no joke. Perhaps on a deeper subconscious level, the brothers are aware that alcohol causes chemical castration. Improved sexual performance may make a man feel like a lucky devil, but hear this, brothers and sisters:

*Alcohol **does** **not** improve sexual performance.*

Female rapper Yo-Yo pitched ST (Satan) Ides to females: "ST Ides puts you in the mood and makes you wanna go OOOOH!" Given the high rates of HIV and AIDS among Black women (and children), the promotion of risky sexual activity is devilish to say the least.

"Alcohol and substance abuse impact the fertility of the African American population."
–Lawford Goddard, Director, Education and Training Institute for the Advanced Study of Black Family Life

Hmmm . . .

What does Ice Cube and the Grim Reaper have in common? Their heads tilt on the same angle and they carry their weapons of destruction in their right hands.

A Poster is Worth a Thousand Words

Let's decode the symbolism and subliminal messages of Ice Cube's infamous poster. We know that alcohol advertisers are aggressive, clever, and deceitful. They utilize every symbol, sign, word, and image to persuade you to buy their product, even to the extent of violating their own industry rules.

The backdrop for the poster is a brick wall. The sly appeal here is to the emotions of hardcore urban Black men. Many inner city Black males feel that their backs are up against the walls of injustice, poverty, racism, etc.

A wall can symbolize isolation, confinement, and a dead end. Many young people who feel confined, isolated, or at a dead end in life are using alcohol to escape. ST (Satan) Ides wants you to believe that alcohol can help you get over the wall. Brothers & Sisters, don't believe the hype.

Ice Cube is dressed like the Grim Reaper, the demonic spirit that comes to claim the soul at the moment of death. We know Black is beautiful, but unfortunately, Western advertisers often play up the color black to indicate danger, evil, and death. In addition to wearing what looks like a hooded black robe, Ice Cube clutches a death symbol in his right hand. He is pointing to the ST (Satan) Ides can with this left hand using a sign that means "I love you" in sign language. Say what! It appears that brother Ice Cube is saying, "I love you, Satan." Of all the dirty tricks we've seen thus far, this one takes the cake. Ice Cube's street credibility was pimped to the max. Remember, the bad saint Satan has the power to make you worship him via deception and deceit.

Now look again at that hand sign. Ice Cube looks like he's flashing his homies a greeting. The Black-P-Stone Bloods of Compton, California flash this sign. Millions of us thought that Ice Cube was flashing the "I love you" sign to a can of ST (Satan) Ides. We've been had.

Notice Ice Cube's eyes on the poster. They appear to be hidden. A dark shadow is cast over his face and eyes, giving the illusion that Ice Cube is wearing a mask.

Within African American culture, young people are encouraged to communicate with their eyes. We tell them to look people in the eye. Our elders know that respect and learning are achieved and enhanced through eye contact, which allows us to absorb wisdom and truth. Also, the eyes can tell us whether a person is being honest.

They say that "The eyes are the windows to the soul," and "The soul is the bed of the truth." In other words, a lie might come out of your mouth, but the truth (or lies) travels from your soul and exits from your eyes. What message is hidden in Ice Cube's shadowed eyes? Who knows? I don't want to believe this brother is consciously saying "I love Satan."

Perhaps another reason for covering Ice Cube's eyes is to make us think he's a criminal. He certainly was in *Boyz N the Hood.* "Oh my God! A gangbanger! I'm about to be killed! He has a gun!"

"What gun?" you may ask. "I don't see a gun in the poster." Examine closely the manner in which Ice Cube grabs the can of ST (Satan) Ides. That can could easily be a gun. Amazing! Ice Cube's index finger lands right on the trigger. This subliminal is absorbed into your subconscious and mixes with the image and hype of the young, Black, violent male.

Brothers & Sisters, why is it that Western society often hides the eyes of its heroes and sheroes with masks? Batman and Robin, the Lone Ranger, Spiderman, Captain America, Batwoman, Lady of Justice, Superman (who hides behind the glasses of Clark Kent), Ku Klux Klan, and the list goes on and on. This is some strange psychology. Why would good people on the side of justice want to hide their eyes (identities)? What do they have to hide?

Maybe their day jobs. Superman, a.k.a. Clark Kent, and Spiderman both work for the mass media (need I say more?). Batman and Robin, a.k.a. Bruce Wayne and Dick Grayson, say they are socially conscious millionaires, but have they helped one family of color? The Lone Ranger is a white man who rides around on a pale horse on stolen land in the name of law and order. He has an indigenous "sidekick" named Tonto. There are many members within the KKK who are politicians, police officers, and judges. I rest my case on Batwoman, the daughter of Gotham City's police chief.

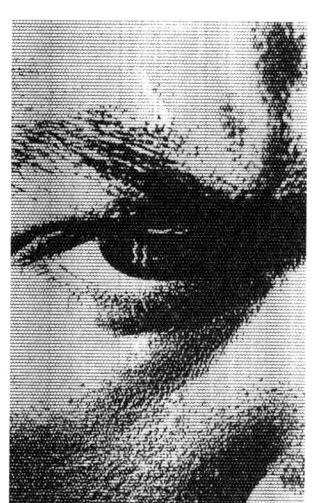

ST (Satan) Ides' rap and hip hop tunes sound revolutionary to young people, but understand this: Ice Cube is pitching for the oppressor, ST (Satan) Ides. Yo, my brothers and sisters in the business, you can't have it both ways. You can't be rapping **REVOLUTION** and **ADDICTION** at the same time. For all you brothers and sisters who pitch alcohol products, consider the words of Paul Lawrence Dunbar: "We wear the mask that grins and lies."

Ice Cube was featured in another ST (Satan) Ides poster that hit the local cop and shop corner stores in the summer of '94. Once again the focus was on Ice Cube's eyes. Partially covered by a shadow, his face conveys angry emotions and evil intentions. In the pupil of his right eye is the blazing red zig zag letter *I* logo. In other words, he's seeing red. According to Webster, red is an angry, hostile, and violent color. A zig zag depicted close to the head indicates hostility, violence, and anger, and remember, the zig zag is a symbol for a snake in the grass. Decoded, this poster reads, "The Black man with the snake eyes is hostile, violent, and evil."

A person depicted as having snake eyes is considered to be a criminal. This poster is supposed to appeal to the so-called hardcore brothers and sisters. When will we wake up? This is their interpretation of us, yet we continue to give them our money and souls.

The Ides Has Arrived

Earlier in the chapter we decoded *Ides* to read "the holy day (holiday) of death." In the summer of '94, a deep TV commercial for ST (Satan) Ides malt liquor aired and starred the hottest rapper in the hip hop business, Snoop Doggy Dogg.

The underlying theme of the commercial was "the Ides has arrived" and that Black people drink it by the truckload. A big black delivery truck is shown cruising through the hood in the middle of the night. The truck is being driven by a Black man dressed in black. His black hat bears the word CROOKED on the front. The brother is so cool, he is wearing sunglasses in the middle of the night. The truck hits a speed bump and a case of ST (Satan) Ides malt liquor falls out the back of the truck onto the ground unbroken. The case lands next to a full grown doberman pincher. The dog's profile slowly transforms into the image of Snoop Doggy Dogg.

The commercial cuts to a house party. Some smooth lyrics with a funky rap is introduced as Snoop walks through the house party passing out 40oz bottles of ST (Satan) Ides from a case.

Singer: *Just hit the corner store* [hit=robbery]
You know what I'm looking for
ST Ides, ST Ides
Snoop rap: *I just come through the door*
with a box of 4-O's [forty ounces]
40s just a bounce and a house full of – [The word was bleeped out, but "hos," i.e., whores, fits. At the exact moment of the bleep, the camera rests on a young Black woman.]

Message 'N A Bottle

> *No sense tripping*
> *Just sippin' on S - T,*
> *That's how we do it in the L-B-C* [Long Beach, CA]
> *ST Ides*

The commercial strongly suggests that the driver of the truck (via the symbolism of being dressed in all black) is a crook (or possibly an undertaker). Furthermore, it consciously reinforces the propaganda that Black men are dogs. When was the last time you saw whites drinking brew by the truckload on TV?

In another commercial, we see what looks like South Central or Compton, California. Several young Black males are being watched from a helicopter (the sound of the helicopter is heard but never seen). Apparently they are under surveillance. The young Black males are shown sitting in front of a house on what appears to be a hot summer day. They're chilling, which reinforces the lazy stereotype of urban Black men. One brother is shown kneeling next to his car, apparently grooming and admiring his tire rims. Suddenly he stands up, walks to the street curb, looks up toward the helicopter, and with one quick action, he reaches behind his back and pulls out a 22oz of ST (Satan) Ides malt liquor. He holds it up to the helicopter and says, "It's at the local corner store." Then the helicopter flies away, perhaps to another ghetto/hood.

This must be TV, because in reality, the only time a helicopter, a.k.a. ghetto bird, hangs over the Black community is when the police is gathering intelligence. The helicopter is the symbol of authority and the person(s) in it represents the hidden power-McKenzie River/G. Heileman- behind ST Ides. The brother's role is clearly that of an informant, or sellout.

I guess everybody's got a price.

This really was TV. If in real life a brother reached behind his back suddenly to pull out anything in the presence of the police, he would have been shot to death on the spot. The 22oz bottle of ST (Satan) Ides was tucked in the back of the brother's belt the same way a gun (a 22 handgun) might be concealed.

Everybody's Got A Price

I've got mixed feelings about those rappers who endorse ST (Satan) Ides and other malt liquors (the list reads like a who's who). Many of them feel like they're caught between a rock and a hard place, and I sympathize. To a point. The real deal is, they're aiding and abetting the spiritual, emotional, mental, and physical murder of our children. Rap artists are idols, so they've got to own up to their responsibility. It's time to do the right thing.

The following is to all those brothers and sisters who refuse to contribute to the real struggle and are willing to push the man's dope in a bottle.

Hmmm...
"You won't hear any more alcohol songs from Snoop Dogg–unless I stumble upon some Hennessy."
–Snoop Doggy Dogg, Playboy, October 1995

You Got Played

They told ya! S-T stood for saint
But a saint it ain't!
You got yanked
S-T is short for Satan
Brew is a spirit so's the spirit of Satan
you've been drinking.
Now you're pissed cause you got played
Black man sober up
It's time to do a bum rush

Rush the truth about the evil brew
They been pimping you

You can't pretend to be blind

Message 'N A Bottle

*to the signs of the times
Now you're pissed cause you got played
You were reckless, should have checked it
Been a skeptic*

*You thought the man was on the level
Didn't know the man had you working for
the DEVIL!*

*Now Satan got you in a bind
cause you signed on the dotted line*

*They played your intellect
wiped your Black pride out like disinfect
Just to earn a paycheck
You put a rope around your own
people's neck!*

*Please, I don't wanna
hear your old excuses
A man got to do
what a man got to do
Satan ain't about the truth
He's about using you*

© 1995 Alfred Powell
and Alfred E. Powell, Jr.

029710217

W A N T E D

NAME: Big Jug

OCCUPATION: Spiritual assassin

ALIASES: BJ, Black Jesus, Jughead

DISTINGUISHING MARKS: Slashed big jugular vein, Christ-like image on cap

VITAL STATISTICS: ht-12", wt-40oz. Cheap: $.89-$.99.

CRIMES: Impersonating an apostle.

AFFILIATIONS: Hudepohl Schoenling of Cincinnati, Ohio. Little is known about this suspect's financial and political connections.

M.O.: Price is cheap so that poor people can afford to buy.

LAST SEEN: In the hands of those looking for Jesus.

CAUTION: Stay clear! Big Jug cannot spiritually cleanse you!

6. Beware the False Prophet

Big Jug, another cheap malt liquor pushed in the Black community, is routinely priced between $.89 and $.99. Big Jug is made by Hudepohl Schoenling in Cincinnati, Ohio and is distributed in Ohio, Kentucky, and Indiana. Like other drugs sold in the hood, its popularity has grown through word of mouth. According to brother Na'im Akbar, pioneer in the development of an African-centered approach to modern psychology:

> We must use our third eye. Our third eye is our spiritual eye, our common sense eye, the eye that allows us to see those things that our physical eyes can't see in the physical world. Our third eye when properly tapped opens up and allows us to see all those things buried deep inside our ancient Nubian memory.[77]

You must open your third eye and tap into your common sense in order to understand the hidden messages and symbols on the label of Big Jug malt liquor. Keep in mind the words of Wilson Bryan Key that "Words are never the things they describe."[78] Since Big Jug malt liquor is pushed in Black communities, we will decode the marketing campaign from an African-centered perspective.

On the surface, the name Big Jug refers to the large size of the 40oz bottle. Maybe Big Jug drinkers think they're getting more malt liquor than from other brands. But a 40oz bottle is a 40oz bottle, no matter what the brand.

When I was a kid playing the dozens, we used to call each other "big jughead." Maybe Big Jug is an abstract of big jughead. In the white community, however, the term big jugs means "big breasts." Words, phrases, and symbols usually have different meanings from culture to culture. Consider how the following words change meaning as they change cultural context:

Word	**Blacks**	**Whites**
crib	home	baby bed
chill	take it easy	cold
bad	good	negative
dope	awesome	illegal drug; stupid person
fly	sharp; good looking	an insect

Jesus and the Jugular Connection

Big Jug has another meaning, and here's where the plot begins to unfold. Did you know that big jug is also medical slang for the **jug**ular vein?

The **big jug**ular vein runs through the side of the neck to the top of your head. It collects blood from the external and internal areas of the skull and face. The skull and face, of course, are right there at brain level. The brain is an amazing piece of work, responsible for thought, emotions, and voluntary and involuntary functions. The brain is the focus of intense study in universities and research labs around the world. The *Chicago Tribune* reported a very interesting study from the University of Illinois Medical School on the roots of violence. Scientists have found that alcohol (a.k.a., 40oz malt liquor) lowers serotonin levels in the brains. Serotonin is one of those feel-good chemicals produced by the pineal gland in the brain that keeps folk chilling, cooled out, mellow. When serotonin levels plunge from alcohol abuse, so go the mellow feelings and inhibitions in some people.[79] When inhibitions are lowered, there is little to prevent you from knocking somebody upside their **big jug**head.

And it gets deeper for Black folks. What they don't tell you in the media is that serotonin polymerizes (chemically changes) into melanin![80] Surely the lowering of serotonin must have a bad impact on melanin production in the body. And who are the melanin people? Black folks!

Since the **big jug**ular vein feeds the brain with nutrient-rich blood, whatever you eat or drink will impact brain functioning. Alcohol first affects the cortex, then the limbic system. The cortex is responsible for reasoning, the limbic system for emotions. Brothers & Sisters, malt liquor use wreaks havoc with your thinking and feeling. Your ability to make good decisions regarding sex, conflict, drug use, etc. is destroyed when you "kill" a bottle of malt liquor. Your feelings become wacked. Is it worth it?

Now answer this riddle: What color is your **big jug**ular vein? Answer: You have *blue* veins, within which flows *red* blood. So what? you may say. Well, carefully examine the front of the Big Jug malt liquor label. A big bold red line runs down the middle of two bold blue lines; all three lines run through the *neck* of the Big Jug 40oz bottle. It appears to be a symbolic representation of your body's own **big jug**ular vein.

The **big jug**ular vein is the lifeline in your body in that it carries blood to your head. Note that the number "40" and the abbreviation "oz." (ounce) are strategically placed over the top of the symbolic lifeline, or **big jug**ular vein. One of the many possible symbolic interpretations of the number 40 is *life*. For example, a woman carries a baby in her womb for 40 weeks. Furthermore, when the bottle is tilted to drink, the "oz" becomes "zo." *Zo* is the Egyptian radical word for life, thus *zo* appearing on top of the symbolic **big jug**ular vein reinforces the idea of a lifeline.

Within the Black community the 40oz bottle has become deadly in many ways. Not only does it contain liquid dope, the bottle itself is often used as a weapon; a smash over the head or a cut to the **big jug**ular vein can cause a person to bleed to death. Notice the symbolic **big jug**ular vein has been cut on the label. There's another way to look at it. Remember that alcohol containers are phallic symbols? The bottle represents the penis (which has a head too) and the brew represents urine which flows out of the penis. If you read the bottle as a phallic symbol, then the cut on the label suggests castration! Think about it (because I don't wanna)!

Whether the **big jug**ular vein has been cut or the penis has been castrated, when a person is bleeding to death, they tend to call on their Maker. Perhaps this is why the image of a crucified Christ-like silhouette (i.e., blacked out) was pictured on the bottle cap.

"Hold it, Coach! What did you say? Jesus on a 40oz bottle?"

That's right–a Christ-like image in silhouette, arms outstretched to either side. We cannot say for sure whether the image is Jesus or not, but we do know this: since the crucifixion of Jesus, the image of a man hanging on a cross has been depicted in many, many ways and in many, many mediums. Whether it is a sophisticated painting of Michaelangelo's uncle hanging on a cross, or two popsickle sticks crudely crisscrossed together, the crucified Christ is the one image that is instantly recognizable throughout the world. As a symbol, it carries many heavy layers of meaning for believers and nonbelievers alike. Just as Jesus Christ is called the spiritual lifeline by millions of believers around the world, the **big jug**ular vein is the biological lifeline in the body and the symbolic lifeline on the Big Jug label.

Christ-like image + big jug**ular vein** = **lifeline**

This claim may be hard for you to believe, so don't take my word for it. From 1992 to 1994, I showed Big Jug's Christ-like image to approximately 1,200 Black high school students from the following cities: Milwaukee, WI, Little Rock, AR, Atlanta, GA, Valdosta, GA, Chicago IL, Winston Salem, NC, Stateville, NC, Cleveland, OH, Washington, DC, and Niagara Falls, NY. I asked them to tell me what the silhouetted image looked like to them. "I thought I saw Jesus Christ. I did see Jesus Christ!" *98% responded Jesus Christ.*

Now compare the Christ-like image on the cap to that of the famous statue of Christ that overlooks the city of Rio de Janeiro in Brazil. The similarities are striking. Both images depict Christ with arms outstretched in the "T" formation. Both are in silhouette.

Big Jug "Christ"

Why did the marketers of Big Jug decide to place a Christ-like silhouetted image on the cap of their bottle of malt liquor? Perhaps Hudepohl Schoenling wasn't aware of Beer Institute guidelines that state religion and brew should never mix in advertising.

Beware of False Prophets

The Bible warns of false prophets, images, and idols.

For many shall come in my name saying, I am Christ; and shall deceive many. (Matthew 24:5)

Rio de Janeiro "Christ"

> For there shall arise false Christs, and false prophets, and shall shew great signs and wonders; insomuch that, if it were possible, they shall deceive the very elect. (Matthew 24:24)

In many ways, Big Jug's Christ-like image has the power of a false prophet. False prophets and alcohol (spirits) can lead you astray, make you feel good, and make you forget your problems. Both are charismatic. Both will deceive you. Whether you are the richest of the rich or the poorest of the poor, the false prophet doesn't discriminate. Brothers & Sisters, the next time you see that Christ-like image on the cap of Big Jug malt liquor, ponder the following: "Beware of false prophets" (St. Matthew 7:15).

Alcohol alters your understanding of truth and righteousness. False prophets twist the truth all out of shape. Perhaps the silhouetted Christ-like image is used to subliminally suggest and persuade the consumers of Big Jug malt liquor that "Christ" (the false prophet on the cap) will protect your liver and kidneys from harm. Brothers & Sisters, you can't worship God and serve the false prophet of alcohol too.

> No man can serve two masters: for either he will hate one, and love the other; or else he will hold to the one, and despise the other. Ye cannot serve God and mammon. (Matthew 6:24)

> I am the Lord thy GOD...Thou shalt have no other gods before me. (Exodus 20:2-3)

> He is God the one and only God. He is eternal, absolute. He begetteth not nor is He begotten and there is none like unto him. (Al-Quaran Ch 112)

The Blood

Earlier we talked about the big bold blue and red line that runs down the middle of the Big Jug label, and we said that the color red symbolized blood. Question: *Whose* blood?

Tertullian said that "The blood of the martyrs is the seed of the church," and he was never lying. Christians base their entire faith on the death and resurrection of Jesus Christ. Scripture says that Jesus's blood brings the believer closer to God, transforms lives, and cleanses all sins (Ephesians 2:13, Hebrews 9:14, 1 Peter 1:19, 1 John 1:7). The notion of his sacrifice is central to the belief. In many Christian churches throughout the world, a ritual celebrating the sacrifice called Holy Communion is held, usually once a month. People drink grape juice (or wine), which symbolizes his blood that was shed, and a wafer (or cracker), which symbolizes his body that was broken on the cross.

We've reported that inside the Big Jug cap is the picture of a Christ-like image in silhouette. Here's the addition:

Christ-like crucifixion image + big red bloodline on the label + liquid dope in a bottle = death and sacrifice

Whose death? Sacrifice of whom? Hang with me, Brothers & Sisters.

The corner where the brothers hang out to drink and talk smack is the "church" and the 40oz bottle in the brown paper bag symbolizes the communion cup. In protestant churches, tiny glasses filled with grape juice are given out to each individual, but in many Catholic churches, one cup is passed from lip to lip. This is exactly what the corner brothers do. They pass the bottle from lip to lip. This is an unholy communion.

"Young blood must have its course, lad, and every dog its day."
–Charles Kingsley

"Blood will tell, but often it tells too much."
–Don Marquis

MESSAGE 'N A BOTTLE

> *"Blood is a cleansing and sanctifying thing, and the nation that regards it as the final horror has lost its manhood...there are many things more horrible than bloodshed, and slavery is one of them!"–Padraic Pearse*

Malt liquor leads to death. Jesus is known as the sacrificial Lamb of God; likewise, our young males in particular are like sacrificial lambs being led to slaughter by the 40oz. And there's more.

Before brothers started calling each other dogg, what did we call each other? Blood! Whose blood is being represented on the Big Jug label? Possibly the blood of Black folks. Blood in this sense symbolizes family and close friendships. Remember as a child you cut your finger and shared blood with your best friend? Blood in this sense symbolizes the *love* we have for one another. With the widespread addiction (love) to the 40oz among African Americans, the blood on the label might be ours. Then,

blood (family/friends) + liquid dope in a bottle + crucifixion theme on the cap = destruction of the Black community

or

Big Jug is the cross that the Black community bears.

And there's more. Remember the vampire? He comes out at night and sucks blood from the **big jug**ular vein on the neck of his victims. Brothers & Sisters, every time you tilt that Big Jug bottle to your lips, you are like a vampire, a bloodsucker, sucking "blood" from the neck of the bottle.

Did you know that when dead bodies are taken to the funeral parlor the mortician embalms the cadaver (dead body) by draining the blood and replacing it with a formaldehyde-based chemical? Brothers & Sisters, drink malt liquor and you'll be like the walking dead, a zombie. Drink Big Jug, or any malt liquor, and the life (symbolically, the blood) will be drained from your body. You'll be spiritually embalmed and

> *"Blood alone moves the wheels of history."*
> *–Benito Mussolini*

ready for the grave. Malt liquor is like the formaldehyde-based chemical that embalms your spirit and your internal organs.

As you can see, blood is loaded with symbolism. In addition, check out the following blood words and terms from the *American Heritage Dictionary*: bloodbath, bloodcurdling, bloodletting, bloodline, menstrual blood, blood poisoning, bloodshed, bloodshot eyes, bloodstain, bloodsucker, bloody mary, cold blooded.

"No one need think that the world can be ruled without blood. The civil sword shall and must be red and bloody."
–Martin Luther

Does It Matter If He's Black or White?

Want to see a black Christ-like image turn white (which is what Michaelangelo did to the real African Jesus)? Stare at the cap image for 33 seconds. (Jesus was crucified at age 33.) Immediately, find a spot on the ceiling, and stare for nine or ten seconds. Watch the black Jesus turn white. You will be amazed.

. . .

Afterthought

Above the Tyler Davidson Fountain in downtown Cincinnati stands a statue of a goddess called the Genius of Waters. She was originally positioned to face East, toward the rising sun. Presented as a gift to the city in 1871, the statue's job is to bless the waters and make them pure. Believe it or not, several miracles have reportedly occurred at its base.

Many of the residents of Cincinnati believe that Big Jug's Christ-like image is the goddess. When the goddess statue is depicted in silhouette, its image becomes remarkably Christ-like. Soon after Shack and I lectured in Cincinnati (May 1993), the Christ-like image was removed from the Big Jug bottle cap. Think the news got back to Hudepohl Schoenling that we were wise to their schemes?

The Genius of Waters

2211212

W A N T E D

NAME: Schlitz Malt Liquor, Red Bull, Bull Ice

OCCUPATION: Spiritual assassin

ALIASES: Forty Dog, Bull Head, Blue Bull

DISTINGUISHING MARKS: Greek letters on the label

VITAL STATISTICS: ht-12", wt-22, 32, and 40oz. 5.8% (blue) and 6.7% (red) alcohol by volume.

CRIMES: Impersonating a Kamitic god

AFFILIATIONS: Stroh Brewery

M.O.: Suspect claims things are back to the way they were and the bull is taking charge. Says he's a "smooth operator."

LAST SEEN: On BET TV *Comedy View*, billboards, and magazines

CAUTION: Suspect will run over you. Stay clear!

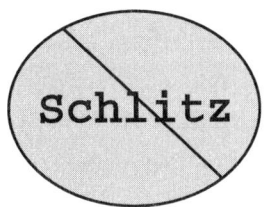

7. The Bull God Osiris Resurrected?

Picasso has been credited with giving modern expression to an ancient idea–that bulls symbolize the essence of the masculine force. In his famous "Half-Man, Half Bull" sketch, Picasso's bull has African features. The sketch looks a lot like pictures of the ancient Egyptian (Kamitic) king-god Ausar, better known as Osiris. Somehow this Spanish painter, sculptor, and ceramist understood that universal power is inherent in the African male. The *American Heritage Dictionary* defines bull as an "exceptionally large, strong and aggressive person. The uncastrated adult male of domestic cattle." Who does that sound like to you (stereotypically, that is)? The Black male!

White males go berserk at this idea. According to Cirlot, "The bull symbolizes the superiority of the Aryan *over* the Negro."[81] Welsing says that's wacked.

> This seems to be a conscious or unconscious attempt by Cirlot to reverse the meaning of the bull symbol since the black bull was the symbol of Osiris, a Black god. Thus, in actuality, the bull must represent the opposite: Black superiority over white in the psyche of the white collective.[82]

Once upon a time in a land called Kamit lived a god-king named Ausar. He was also called Lord of the Perfect Black and the god of fertility. His symbols were the bull, sun, and moon.

Now King Ausar had an evil brother named Seth. Seth was very jealous of Ausar. He hated Ausar so much that he conspired, with 72 other people, to kill him.

Seth and the gang decided to throw a big party, full of wine, women, and song. It was the social event of the season, and King Ausar was the "honored" guest.

Message 'n a Bottle

Just when the party started to heat up, Seth unveiled a pile of treasures which included, of all things, a casket. Ausar was a rich god-king, but still, the sight of all those treasures excited him. Seth smiled wickedly to his brother and said, "He who can fit his body into the coffin will be rewarded with all these treasures."

His decision making ability severely impaired by the ancient 40oz, King Ausar stumbled into the casket and lay down. Quickly, the ungodly gang sealed the coffin shut with silver molten lead. They threw the coffin into the Nile River and left Ausar to his fate.

Sometime later, Ausar's beautiful sister/wife Isis found the coffin and dragged it out of the Nile. The exercise left her exhausted, so she lay upon it and fell alseep. As fate would have it, Seth just happened to be in the hood, and he was enraged to see that the coffin had been rescued.

By the light of the full moon, Seth took Ausar's body out of the coffin,

Do you think Stroh Brewery Co. of Detroit, Michigan, is aware of Cirlot's take on the bull? It's an interesting question, given the fact that their bull products–Schlitz Malt Liquor, Red Bull, and Bull Ice–are exclusively marketed towards the Black community and specifically pushed into the culture of young Black men.

Stroh's vice president Michael Porter said that "the bull has been a part of the Schlitz Malt Liquor imagery from its conception. It was to differentiate the product from beer, to tell people this product is marginally stronger than beer."[83] In other words, the bull is a symbol of strength and might.

Brothers & Sisters, don't be fooled. Advertisers are psychologists with marketing skills. They fully understand human motivation, symbolism, and ethnic culture. While Picasso honored the original symbolic meaning of the bull as the African male force, Stroh's blue-black bull ("crystal-black,"* as Welsing says) serves as a subliminal reminder of the alcohol industry's domination over the Black community. Perhaps Cirlot's interpretation guides Stroh's marketers; they have seen fit to twist the ancient African myth of Osiris to suit their needs. Alcohol companies and their advertisers often feel free to take liberties with a people's myths, symbols, and archetypes–as do Western scientists, historians, theologians, anthropologists, archeologists, historians, and artists. For example, Red Wolf and Crazy Horse malt liquors exploit sacred Native American symbols and history.

———————

*In Black culture, "crystal black" means blue-black.

SCHLITZ MALT LIQUOR

The Cult of the Bull

Picasso's depiction of a bull with African features had deep cultural meaning for him. Picasso was a Spaniard. Spanish men have an unusual love-death relationship with bulls, and it manifests in two important cultural events. *La Corrida de Toros* ("the run of the bull") is the annual rite of passage event for men held in Spain every year on the seventh day of the seventh month at exactly seven a.m. (3 x 7 = 21. Twenty-one is the age of manhood for Europeans.) In *La Corrida*, a herd of bulls chase mostly Caucasian men (e.g., American writer Ernest Hemingway participated once), down narrow, winding streets.

One commercial for Schlitz Malt Liquor cleverly uses *La Corrida de Toros*. We see a Black man hurrying to his local corner store to buy a bottle of Schlitz Malt Liquor. Instead of the bull chasing him, he's chasing the bull, i.e., the alcohol. At the end of the run, the largest and strongest bull is lead into a bullring (circle of death), where the beast is put to death. The moral of the story is: as long as you chase the bull (alcohol), you will be put to death.

The other popular bull event is the bullfight. Given the symbology of the bull as the African male force, this sport may symbolize the ritual celebration of Europe's military defeat of the African Moors in 1492. (The Moors had ruled Spain for seven hundred years.) In Spain, bullfighting is as big a deal as the Superbowl is in America. Stadiums are packed with people who love to see matadors (i.e., bullfighters) dressed in fancy costumes, this time chasing the bull around the ring. The contest is won when the matador stabs the bull in the back! As Dr. Welsing says, "If the bull was the symbol of Aryan superiority over the Negro, there would be no need for whites to kill a black bull with a sword."[84] Cirlot's reversal doesn't stand up.

chopped it up into 14 pieces, and scattered them about the earth. Isis managed to retrieve 13 pieces (she couldn't find the penis). Then she put Ausar back together, prayed for a child, and immaculately (i.e., without sex) conceived a baby boy she named Horus. (By the way, many scholars believe that the story of the immaculate conception of Jesus was taken from the Ausar myth, but that's another story.) Horus eventually grew up to avenge his father's death by killing Seth. And the resurrected Ausar became Lord of the Underworld, Judge of the Dead.

Message 'n a Bottle

The victorious matador cuts up the bull and offers its ears and tail as souvenirs. The bull's testicles are often eaten as a delicacy. Is it mere coincidence that Osiris, the Egyptian (Kemetic) god whose personal symbol is the bull, was also cut up into many pieces by his evil brother Seth?

In the Osiris myth, the god's body parts are scattered to the ends of the earth. Decoded this means that bulls (i.e., African males) are at risk all over the world of dismemberment. For example, in America at the turn of the century, Black men were routinely lynched in the South. Often, the black penis would be cut off from the corpse, stuffed into a pickle jar, and prominently displayed in the window of the small town mom and pop grocery store.

Jackie Sherrill, coach of the Mississippi State football team, knew what he was doing when, on September 3, 1992, he supervised the castration of a bull named Wild Willie (Horton?) in the presence of his mostly all Black male team. The poor bull didn't even have the benefit of an anesthetic to deaden the pain. Sherrill says he was trying to motivate his players to beat the Texas Longhorn team whose star players were big aggressive Black men. The bull's castration also may have served to threaten his own big Black players: "Win, or you'll be *cut* (castrated) from the team." At any rate, the stunt worked. Mississippi State won 28 to 10.[85]

Castrating the Black man as well as the black bull is the symbolic, ritual attempt of white men to capture the real power of Osiris, the Black Kamitic god whose death involved castration. By the way, the bulls' genitals on the labels of Schlitz Malt Liquor and Red Bull are missing. Where'd they go? The slogan for Red Bull and Schlitz Malt Liquor is, "It's the real power" and "No one does it like the bull!" Decoded, the new slogan reads, "No one does it like the bull god Osiris."

SCHLITZ MALT LIQUOR

Lots of Bull

Since the latter part of the 14th century, the European has redefined the symbolism of "bull" and "black," ripping their sacred and spiritual meanings from the pages of African history. Both terms are now full of negative imagery and connotations. They cause instant panic on behalf of white supremacy. Consider the following terms, code phrases, and symbols.

If you drink too much alcohol, you will fall down and break your leg.
–ancient Kamitic saying

> *Variations on the bull theme:* Bull Conors, bullsh-, bullwhip, bully, bullies, bull dog, pit bull, bullet, bull rat, bulldagger, bull headed, bull balls, bullish, bullfighter, bullfighting, bullring, bulls eye, "shoot the bull," "that's bull," bull rush, bulldozer, bulls lips, bull dick, bull neck, big black bull, mean as a bull, Red Bull, Schlitz Malt Liquor Bull, Bull on Ice.
>
> *Variations on the black theme*: blackball, blackmail, Black Sunday, Black Monday, Black Death, blacklist, black mark, black magic, black comedy, black nigger, black market, blackout, Black men, big Black men, black lie, black sheep, black widow, black bird, black head, black face, black jack, Black Flag, black frost, Black Guard, black funeral, black ass, black clouds, black day, black liar, black bag job, blackboard jungle, black book, Black Plague, black leg, black project.

Decoded, the Schlitz con goes something like this:

If crystal-black bull = Osiris/African manhood, then

crystal-black bull + Stroh's targeting of young Black males + missing genitals on the Schlitz bull + 40oz of liquid death =

> *Chemically castrate the big virile Black male!*

"One evening in a crowded bar, I heard the familiar noise of a bottle shattering amid the patrons," said Harvey J. Brown, president of Get a Grip II Inc. "I thought to myself, there's a real consumer need here– a package that's easy to hold, and to grip securely...". (The thoughts behind the creation of Stroh's new 16oz Gripper, which in street language refers to the enforcer, the gun man in a gang.)
–Milwaukee Journal Sentinel, *April 15, 1995*

Furthermore, have you noticed that the tail of the bull on the label is shown erect? A bull's tail will become erect on two occasions: during intercourse and when it's defecating. Now if the bull on the label is missing its genitals, you tell me why its tail is erect!

As we've mentioned before, alcohol does not make your "jimmy" thicker. It can literally kill the sexual response in a male. Alcohol can make a man impotent. That's what the crystal-black bull without testicles on the Schlitz label represents–the chemical castration of the Black male.

Indiana University basketball coach Bobby "The General" Knight found it humorous to use a bullwhip to "motivate" his basketball team. (Believe it or not, this man actually received a degree in history from Ohio State University.) Imagine this scene straight out of America's slavery past: Black superstar Calbert Cheney, shorts slightly sagging, is being held down by Assistant Coach Norm Ellenberger while Coach Knight (the historian) playfully whips Cheney's rear end with the bullwhip.[86] Coach Knight must have been sleeping in history class the day when the brutality of slavery under the bullwhip was being discussed. On second thought, maybe he was wide awake.

It's clear that brother Cheney does not know his own history. Surely he would not have knowingly participated in such a stunt that must have had his own ancestors rolling around in their graves. The bullwhip is a very powerful symbol in the collective psyche of Black people, and before any Black athlete joins a team at Indiana, he or she should demand that Coach Knight and his staff read *Bullwhip Days: The Slaves Remember* by James Mellon. Or maybe they should talk to Kunta Kente, who was renamed Toby under the torture of the bullwhip.

And finally, Brothers & Sisters, consider for a moment

the terminology used in the sports world. Commentators routinely refer to Black athletes as bulls. Black boxers are praised for charging into the ring like a bull. In a basketball game, towering Black men are referred to as bulls under the basket. Who is the most famous Bull of all time? Michael Jordan, a Black man! In football, big Black running fullbacks are cheered for running like bulls.

Comparisons of the Schlitz Malt Liquor Ad Campaign to the Osiris Myth

Stroh Brewery was able to pull a fast one on the Black community because we don't know our own history. The parallels between the Schlitz ad campaign and the Osiris myth will amaze you.

1. In most Schlitz ads, the moon is a prominent image. "In *The Book of the Dead*, the scribe Ani says to Osiris: Hail, One, rising (shining) from the Moon! Hail One, Shining from the Moon. Osiris, the Moon-God."[87]

2. In a TV commercial for Schlitz Malt Liquor, a Black man is shown lifting weights (virility). Later he is seen romancing a woman (fertility). Clearly, the two actors represent Osiris and Isis, who were brother and sister as well as husband and wife. At the end of the commercial, the man is shown walking down the street under the light of a full moon. *His shadow is projected onto a wall as a big black bull.* In the "Smooth Operator" print ad, the couple is shown dancing under the light of the full moon. A full moon can mean romance, or it can mean sexual violence! Alcohol has been linked to 52% of rapes and up to 50% of spouse abuses cases in the United States.[88]

In the *USA Weekend* write-in [survey with 93,000 student respondents], 47 percent of 10th- to 12th-grade boys say they drink; 28 percent of girls that age report drinking. Most girls say alcohol makes young people more violent or aggressive; boys are more likely to view drunken behavior as fun or relaxing. What many boys consider an escape from reality, many girls consider a loss of control.[89]

Is it any wonder that such violence was done to Osiris under the light of the full moon?

3. Black culture identifies the smooth operator as a player, pimp, mack daddy, and hustler. Osiris's evil brother Seth was depicted as wicked, artful, crafty, slick, charming, clever, and cunning–or a *smooth operator*. Coincidence?

4. Two titles bestowed on Osiris after his death was "Judge of the Dead" and "Ruler of the Underworld." Ruler of the Underworld can be decoded to mean "Ruler of the Night (or Night Life)." Western culture resonates to the shadow side of Osiris's personality. Thus the imagery of the underworld is about chaos, evil, death, and loose behavior. In a TV commercial for Schlitz Malt Liquor, a big black bull (Osiris) leaps from a shining full moon and trots across a dark (night) skyline. In Western culture, the stereotypes of night life are synonymous with the evils of the underworld. Also, the bull leaping from a shining full moon is consistent with the Osiris myth. In *The Book of the Dead*, the scribe Ani says to Osiris, "Hail, One, rising (shining) from the Moon!"

5. In another Schlitz TV commercial, a Black man is seen holding a bowling ball. Usually bowling balls are black, however, this bowling ball is dark blue and resembles pictures of Mother Earth taken from outer space. The brother rolls the ball down the lane. In the real world of bowling, a big black ball is rolled down a lane in an attempt to knock down ten *white* pins with *red* necks. (Dr.

SCHLITZ MALT LIQUOR

Welsing has a field day with the racial/color clues in this ball game.) In the commercial, however, the pins are darkly colored. Clearly, the dark pins are phallic symbols of Osiris, and the ball represents Mother Earth. When the ball hits the phallic pins, the scattering of the body of Osiris about the Earth is reenacted.

6. According to mythology, Osiris was castrated (operated on) by his evil brother Seth (the smooth operator). We've argued that the bull without genitals on the Schlitz label is symbolic of the castrated Osiris (and Black male).

But wait a minute. Could it be that Osiris's genitals have been staring us in the face all along, so to speak? If you want to hide something, make it obvious. Put it right out in front. Examine closely the artificial inseminator, i.e., the 40oz bottle. Both penis and bottle are dark and long, and both contain potent fluids. Think about it. Every time you drink from a 40oz bottle of Schlitz Malt Liquor, you're drinking from the penis of Osiris! Not you, because you pour yours into a glass? Question: what color is the fluid that comes out of the 40oz bottle? Golden brown! What color is the fluid that flows from the bull's penis? Golden brown! Pucker up, all you Schlitz Malt Liquor 40oz drinkers.

7. Red Bull is also consistent with the Osiris myth. In the "Awakening Osiris" chapter of *The Book of the Dead*, Ani writes of Osiris, "I am the **Red Bull**-calf which is marked with markings. The Gods shall say when they hear of me: Uncover your faces."[90] In ancient times, red bulls or calves were often sacrificed, i.e., killed! And isn't it interesting that the dead body of Osiris is frequently depicted as red.

In the summer of '94, several 30 second Schlitz Bull commercials were released. One featured a Black man

standing in the middle of a bullring (circle of death) waving a red cape. The doors to the bullring shake violently when suddenly they fling open and a 40oz bottle comes charging in. The slogan, which is introduced at the end of the commercial, goes, "The bull is taking charge." Clearly, the bull represents the Black man, the red cape symbolizes agitation and bait, and the matador is the symbol of authority. When the bull (Black man) is seduced into charging the symbol of authority, the bull is put to death. Stroh is conning the Black male with the slogan, "The bull is taking charge." You may think that alcohol can help you take charge, but of course that's a lie. How can you take charge if you no longer have control over your thoughts, feelings, and behavior?

8. Seth was jealous of his brother Osiris, so he plotted to kill him. Seth invited Osiris to a banquet in his honor and tricked him into lying down into a beautiful coffin. When Osiris foolishly took the bait, Seth and his 72 conspirators nailed the coffin shut and sealed it with molten lead. Molten lead is silver, the same color as the Schiltz Malt Liquor can. Every time you drink from the silver can, you're symbolically drinking from the casket of Osiris.

Bull On Ice

The most important characters in Egyptian mythology are Osiris, Isis, and Horus. Many scholars equate them to God, Jesus, and Mary in Christian theology.[91] Amazingly, the Bull Ice malt liquor label is profoundly consistent with this line of thought. Consider the following breakdown:

Bull Ice Label		Egyptian Mythology		Christian Theology
Bull (symbol)	=	Osiris	=	God

132

Bull Ice Label	Egyptian Mythology	Christian Theology
Rising Sun = (symbol)	**Horus** = (the son of a god, born of a virgin)	**Jesus** (the Son of God, born of a virgin
Ice = (symbolic virgin)	**Isis** = (a virgin)	**Mary** (a virgin)

When comparing the lives of Jesus and Horus, we learn that both were born of a virgin; both were baptized at the age of 30; both shared December 25 as a birthdate; and both are referred to as the "Son of God." Osiris battled Seth (also Set or Seti), and Jesus battles Satan. Both were violently killed and then resurrected, and both are characterized as the judge of humanity.[92]

The advertising scheme for Bull Ice takes a different angle with the Osiris myth. It uses the symbol of a rising sun rather than a full moon. Scholars agree that one of the many symbols of Osiris is a rising sun. As we begin to examine the Bull Ice label carefully, the obvious and not so obvious links and parallels to the Osiris myth will begin to "shine" through.

- The term "ice" is alcohol industry jargon that refers to a fettering process that makes ice beers higher in alcohol content than regular beers. In the night street culture into which Bull Ice is being pushed, however, the term means "death." Ice is also the street name for an extremely dangerous illegal drug. Thus, Bull Ice can be decoded to read "the Black man's death" or "Osiris's death."

- A sun is used as the backdrop in the Bull Ice ad. Why use the sun symbol after years of using the moon? The sun symbol can be interpreted many ways within the context of the Osiris myth. Within the Greek version of the myth, the sun and moon are the eyes of heaven; the sun is the symbol of good, the moon, the symbol of evil. Osiris is depicted as both sun and moon, good and evil.

- The sun (a light giver), which is above and behind the head of the bull (Osiris), connotes that the bull is a light bearer. The literal meaning of Lucifer, the biblical devil, is "bearer or bringer of light." After Osiris disappeared from the sky, the ancients began to understand him as the representation of Lucifer. Many secret societies base their initiation rituals on Osiris's connection to the sun: "How art thou fallen from heaven, O Lucifer" (Isaiah 14:12).
- The setting (dying) sun is symbolic of the death of Osiris. Thus the bull on the label can be decoded as a corpse. Once again this parallels the myth of Osiris. There are Egyptian paintings of a black bull bearing the corpse of Osiris on its back. The setting (dying) sun can also symbolize a fall from grace in the Luciferian sense, which is what happens to people who get addicted to malt liquor.
- Note that only the head of the bull is shown. It's as if the *body* has been chopped off. This dismemberment is consistent with Osiris being chopped into fourteen pieces.
- When you first look at a bottle of Bull Ice, you probably wouldn't notice the bull on the inside of the label. It is literally hidden. According to Gerald Massey, Osiris was also called the "Hidden-Face."[93]

The Stone and Greek Letters

Look again at the silver can. The seal that appears on the label of Schlitz Malt Liquor provides the backdrop for the large crystal-black bull (Osiris). Notice that the seal, which is round and gray, looks like stone. This idea is further reinforced by the Greek letters that are chiseled there somewhat imperfectly.

According to many scholars and Egyptologists, many of the miracles associated with the life of Jesus Christ were borrowed from the Osiris myth. If the silver can represents Osiris's casket that was sealed by silver molten lead, then perhaps the seal/stone on the label represents the stone that was rolled away from the tomb of the biblical Jesus Christ.

> And when Joseph had taken the body, he wrapped it in a clean linen cloth, And laid it in his own new tomb, which he had hewn [chiseled] out in the rock: and he rolled a great stone to the door of the sepulchre [tomb], and departed. (Matthew 27:59-60)

We have argued that the sun on the label is a setting (dying) sun, but scripture suggests that it could be rising.

> And very early in the morning the first day of the week, they came unto the sepulchre [tomb] at the *rising of the sun*. And they said among themselves, Who shall roll us away the stone from the door of the sepulchre [tomb]? And when they looked, they saw that the stone was rolled away: for it was very great. (Mark 16: 2-4)

When it was discovered that Jesus's body was gone, the believers knew that a resurrection had occurred. Perhaps that's why the seal/stone is so dominant on the label. The seal/stone could very well symbolize a tomb. The depiction of the bull (Osiris) in front of, or *outside of*, the stone (tomb) tells us that the stone (tomb) is empty. A resurrection has occurred.

> So that the people could realize that Osiris had conquered the grave, it was necessary for the tomb to be empty. It was the only way to impress them with the fact that the Lord had risen.[94]

The Egyptian, Christian, and Schlitz parallels are amazing!

Message 'N A Bottle

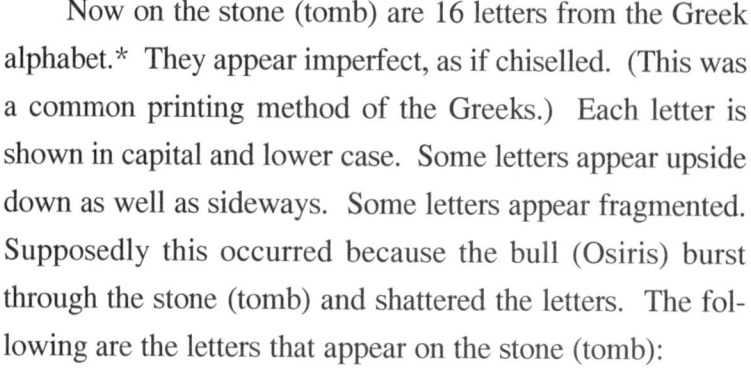

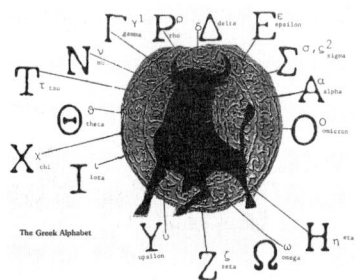

Now on the stone (tomb) are 16 letters from the Greek alphabet.* They appear imperfect, as if chiselled. (This was a common printing method of the Greeks.) Each letter is shown in capital and lower case. Some letters appear upside down as well as sideways. Some letters appear fragmented. Supposedly this occurred because the bull (Osiris) burst through the stone (tomb) and shattered the letters. The following are the letters that appear on the stone (tomb):

Small Letters	Capital Letters	Name
α	A	alpha
δ	Δ	delta
ε	E	episilon
γ	Γ	gamma
η	H	eta
ι	I	iota
ν	N	nu
ο	O	omicron
θ	Θ	theta
ρ	P	rho
σ	Σ	sigma
τ	T	tau
υ	Y	upsilon
ζ	Z	zeta
ξ	Ξ	chi
ω	Ω	omega

There are 24 letters in the Greek alphabet, but Stroh only put these 16 on the stone (tomb). Why these particular letters? If you play the letter game, you'll find that the 16 Greek letters spell out many interesting words. (Again, the letters were "chiselled" and some are difficult to read. These are the

---

*People often ask how I'm able to see the Greek letters on the stone. The bulls on the can and bottle are so big, some of the letters are covered up. Actually, the bull on the box is smaller, thus the letters are revealed for everyone to see.

SCHLITZ MALT LIQUOR

Greek spellings and come pretty close to the rough letters on the stone (tomb).)

Label	Greek	English
ιΗΣΟΣ	Ιησουζ	Jesus
ΞΡιΣΤΟαΣ	ΞΡΙΣΤΟαΣ	Christ
υιοΣ	ΥιΟζ	Son
θΕΟυ	θ*E*Ου	God
Σω*TηP*	ΣωΤΗρ	Savior

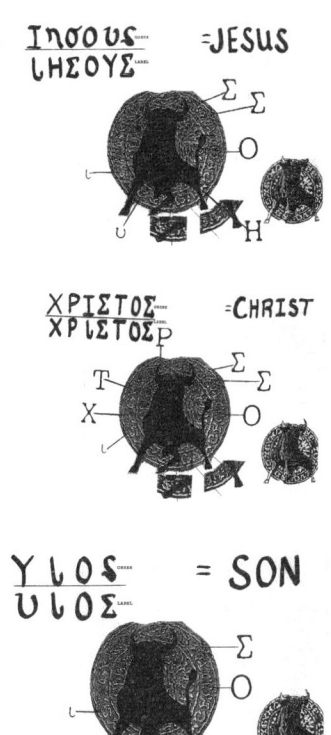

Decoded, the stone (tomb) on the label suggests that alcohol can be your salvation. Brothers & Sisters,

"Keep sane and sober." (1 Peter 4:7)

"Be sober, be watchful. Your adversary the devil prowls around like a roaring lion, seeking someone to devour." (1 Peter 5:8)

"Don't believe the hype."–Public Enemy

. . .

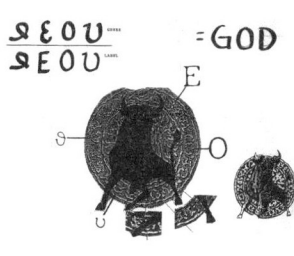

Carefully examine the word *Schlitz*. Notice the word *schiz* inside. According to Webster, schiz is short for schizoid, schizophrene, schizzy, and schizophrenia, all of which means "separation of the intellect from emotions, withdrawal from social relationships, delusions, and hallucinations. Distortions in normal logical thought processes." How many times have you said, "He acts like a different person whenever he drinks alcohol." That Schlitz can make you schizoid. The slogan, "No one does it like the bull" should read, "No one makes you schiz like the bull!"

There's another word inside Schlitz. It is "sh-." **Bull + sh- = bullsh-.** Yes, symbolically you've been drinking bullsh-. Shocking!

137

002-22-4

W A N T E D

NAME: Budweiser

OCCUPATION: Supremacist

ALIASES: Bud, the King of Beers

DISTINGUISHING MARKS: Red, white and blue color scheme. Eagle flying westward through the "A" with three arrows in its talons (which can either appear broken or unbroken).

NATIONALITY: American (German ancestry)

VITAL STATISTICS: ht-12"; wt-22, 32, and 40oz; 4.7% per volume.

CRIMES: Promotes racism via symbols.

AFFILIATIONS: Anheuser-Busch Companies Inc.

MISSION: To promote white supremacy; to teach African history to Black people.

M.O.: "Proud to be your Bud," "It's a Bud thing," "No one beats a Bud," "When you say Budweiser, you've said it all."

LAST SEEN: Suspect spotted on national TV promoting a prize fight which pitted men of color against each other.

CAUTION: Suspect is financially powerful.

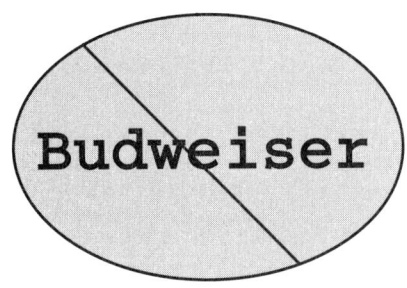

8. The Final Confrontation

> If you do not understand white supremacy (racism)–
> what it is, and how it works–everything else that
> you understand, will only confuse you.
> –Neely Fuller, Jr.[95]

I know Budweiser isn't a malt liquor, but people who attend our lectures always want us to decode Budweiser beer from an African-centered perspective. So we start with Neely Fuller's statement about white supremacy, and from that idea, Anheuser-Busch's marketing scheme begins to make sense.

In order to fully understand the hidden message found on the Budweiser label, I found it helpful to look at Jung's perspective on symbols.

> A symbol is a term, a name, or even a picture that may be familiar in daily life, yet that possesses specific connotations in addition to its conventional and obvious meaning. It implies something vague, unknown or hidden from us.[96]

We may *see* an image or read a word or a phrase, but we *think* and *feel* on many, many levels. That's what is meant by the saying, "One picture is worth a thousand words."

Hmmmm...
The folks at Budweiser find it necessary to teach Black folks about their own African history, so they produced a series of posters depicting African kings and queens, thus linking our ancestors' greatness to their alcohol products. I bet they'll never produce a series of posters featuring popes, European kings and queens, or American presidents.

139

Budweiser's color scheme is red, white, and blue. Both the American and British flags are red, white, and blue. This is a deep color scheme that has entered into the realm of the symbolic. This color combo has the power to evoke feelings of patriotism for many Americans and Europeans. On the other hand, many people of color around the world react to the red, white, and blue with fear and/or hatred. For them, it is a symbol of Western white supremacy. According to Dr. Welsing, the red-white-blue color scheme stands for the **red** blood which runs through **blue** veins which can be seen under **white** skin.[97] Stars and crosses are other prominent symbols on American and European flags that reinforce the notion of supremacy.

For years the slogan "No one beats a Bud" was the very popular tagline for Budweiser beer. Now I'm a coach, and a successful one at that, so I had to ask myself the question: What is this game that Budweiser plays and wins every time? As I studied the Budweiser label, I began to realize that Budweiser's game is racism.

The Diamond

The diamond is a prominent image on the Budweiser label. What's up with that? Well, from an African-centered perspective, the richest source of diamonds in the world is Africa, but let's get even deeper than that. Budweiser is the leading TV sponsor of baseball in America. It just so happens that the Budweiser label bears a diamond that is centered inside the circle that is located above the word *Budweiser*. Clearly, this is a symbolic baseball diamond, and it is surrounded by the following words: America, Asia, Europe, Australia, and Africa.

Positioning on the symbolic baseball diamond is important. Behind home plate (the hind catcher position) is America.

In the game of baseball, the hind catcher is the field general who decides what pitch the pitcher will throw. He must also protect home plate by any means necessary. The hind catcher position connotes superiority and supremacy. The catcher is the only player on the field dressed for "war." The only threat to the hind catcher is the runner on third base.

Who's on third? Africa! If you don't believe me, go check it out for yourself. The symbolic message here is as clear as night and day, or better yet black and white. If the runner on third base is African (i.e., a Black man) and the hind catcher is American (i.e., a white man), then the final confrontation at home plate will be Black vs. white. Baseball may be known as the great American pastime, but the truth is, *racism* is. It's like hot dogs and apple pie. It's a symbolic irony that those great players who mastered the art of stealing home plate have been African American. The first Black player who had to play the game of baseball and racism simultaneously was Jackie Robinson.

For years the bold tagline for Budweiser beer was "No one beats a Bud." Perhaps the slogan is supposed to remind us that Bud is a member (a "genuine" member, according to the label) of an institution that is unbeatable. The tagline is profoundly consistent with the myth that the institution of racism is unbeatable, but don't believe the hype. The slogan "No one beats a Bud" can be decoded to read "No one beats a buddy system," a.k.a., the Good Old Boys Club.

People often ask me to explain why I assume that Bud represents Euro-white males. First of all, how many brothers do you know in the hood named Bud or Buddy? Probably not many. Secondly, Anheuser-Busch told us themselves when they created the "Bud Man" ad campaign. Bud Man was a white male "superhero" who sported the red, white, and blue color scheme and a mask. In chapter 5 we discussed

"Deprive a people of its history and you deprive them of both meaning and identity."
–Dr. Martin Luther King, Jr.

the unusual phenomenon of white superheroes hiding their identities behind masks. Given the fact that most criminals wear masks, what crimes must Bud Man be hiding behind his?

What's In A Name?

The name "Bud" or "Budd" is an olde English term that means *messenger*. The name "Weis" or "Weiser" means *albino-haired man*.[98] A person with albino hair usually has albino, i.e., white, skin. Thus,

Bud + weiser = messenger with white skin

Or, as brother James Anyike says, "In an African-centered context, the name Budweiser means 'a messenger with little or no melanin'."

Name That Tune – *Bud*

There is a Budweiser ad that is very popular in the Black community. You see it on billboards throughout Black neighborhoods and in Black magazines, e.g., *Ebony* and *Jet*. The billboard and print ad in question reads, "dig this, it's a Bud

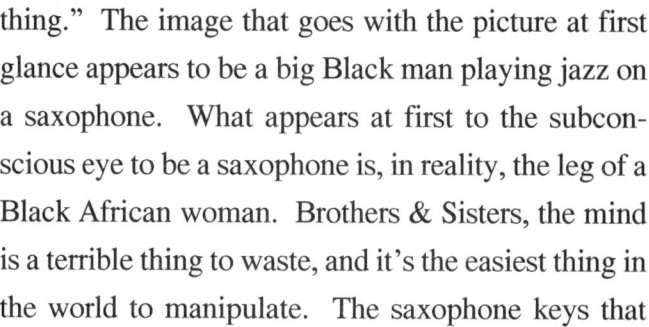

thing." The image that goes with the picture at first glance appears to be a big Black man playing jazz on a saxophone. What appears at first to the subconscious eye to be a saxophone is, in reality, the leg of a Black African woman. Brothers & Sisters, the mind is a terrible thing to waste, and it's the easiest thing in the world to manipulate. The saxophone keys that run down the leg of the sister are similar to the ankle bracelets commonly worn within African and African American cultures.

Among musicians, the saxophone is considered to be a sensual instrument. Together, artist and sax make sweet music. Some jazz artists refer to their saxophones as "sexophones."

Clearly, this ad is about sex, and the musician appears to be "playing" the leg of a dark-skinned woman. Note that each toe is stretched and erect. The slightly opened lips of the jazz man are reaching out to touch them. Also, the woman to whom the leg belongs appears to be lying on her back, a very vulnerable position, while the big Black man is towering over her. Her semi-bare leg suggests that she is naked as well. The man is so big, and the leg is so much smaller that the name of this tune could very well be rape or even child molestation. The artwork promotes the stereotype that Black men are sexual predators.

The slogan, "It's a Bud thing," plays on the Black cultural jargon, "It's a Black thang." And many Black jazz musicians still say "dig this" (i.e., "understand this" or "check this out"), even though that phrase was more popular in the '60s. When you combine the slogan with the picture, you get the following decoded messages:

1. Dig this, it's a Black thing for Black men to play (play around on) their Black African women.
2. Dig this, it's a Black thing for big Black men to molest young Black girls.

Cultural Sabotage

By now, most TV watchers have seen the infamous "Black Ant" Budweiser commercial. Bob Garfield of *Advertising Age* magazine gave it three and one-half stars out of a possible four.[99] It is important that Black folks decode this commercial from an African world view. This commercial is not innocent entertainment. We must study and interpret what is taking place in the TV commercial from a position of truth and correctness, faith and justice, *his*tory and *our*story. With a quickness we must inform every African throughout the diaspora about Anheuser-Busch's depiction of African people via symbolism.

Brothers & Sisters, whether we realize it or not, our fate is linked to the fate of Mother Africa and other Africans around the world, thus it is our duty to correct the negative imagery that Euromedia pumps into the mass mind about our people and the land of our ancestors.

As I attempt to deprogram people, I routinely ask my audiences to give their interpretation of the "Black Ant" commercial. Most people find the commercial funny. The idea of big black ants carrying a 22oz bottle of Budweiser across a desert, placing it inside an anthill, then jamming to K.C. and the Sunshine Band ("Do a little dance/make a little love/get down tonight/get down tonight") cracks folks up.

Most people, especially Black folks, don't realize that the commercial makes a joke out of the historical suffering of Africans and Mother Africa. Ignorance is rampant and deadly in our community. If I play the commercial once, nine times out of ten I'll get requests for encores. The more I play the commercial, the more the audience laughs–until we begin to really look at what the commercial is saying.

At the beginning of the commercial, we hear the beats of the tribal drum and the rhythmic African chanting. Question: On what continent is this activity taking place? Answer: Africa, Mother Africa, the Motherland, the Cradle of Civilization. What part of Africa?

Carefully reexamine the terrain in the commercial. Look at the dry bushes in the background and the dead dry tree in the foreground. **Dry terrain + dry bushes and tree = severe drought and famine.** These clues lead us to countries

"We didn't realize that as soon as we were made to hate Africa and Africans, we also hated ourselves."
–Malcolm X

BUDWEISER

in northeast Africa, possibly Somalia, Ethiopia, Sudan, Kenya, or Egypt. We could also include South Africa (Azania). Famine and drought are ongoing problems in these parts of Africa. The severely dry terrain in the commercial strongly suggests that Mother (Earth) Africa is dying of thirst, that she is in need of rain water.

The early drumming and chanting is profoundly consistent with the African cultural ritual of praying for rain. Coincidence? Furthermore, rain water is needed to sustain life, and water is a symbol for life. Simply put, no rain water, no life! When a country suffers severe famine and drought, crop damage, malnutrition, dwindling food supplies, war, dying livestock, disease, and death result. Many negropeans and middle class Black people have a false sense of security as they live large in America, but they fail to see that their fate is tied to the condition of our brothers and sisters on the continent. Mother Africa is dying of thirst while here in America, her sons and daughters are drowning in the 40oz. There's more than one way to kill a big black ant, as we will soon see.

The commercial shows a scout ant, apparently looking for a special spot on the land. He stops and whistles to the other black ants to come on. Clearly this scout ant is in charge over the other black ants. In African American culture we would say he's the overseer for "the man," the "head ant (nigger) in charge." I find no humor in this at all. In fact, it's rather insulting.

Notice that there are no humans in the commercial, only big black ants. Question: Who do the big black ants walking across the continent of Africa represent? Answer: Africans! Black males! Now the big black ants (Africans/Black males) are

carrying a glistening (wet) 22oz bottle of Budweiser beer across the terrain. Budweiser is known as the King of Beers! The King! Oh, damn! That's right, just like in the Tarzan movies, the big black ants (Africans/Black males) are carrying the king across the African terrain.

Question: Which country does this symbolic king come from? Answer: Europe! We know he's from Europe because Budweiser is a European name.

Question: What do big black ants and African people have in common? Answers: (1) a value system that stresses cooperation, collective responsibility, and interdependence. (2) Ants live in colonies, and African people were colonized by the Europeans. (3) Like ants, African people have been stepped on spiritually, psychologically, economically, and physically.

The barbed wire running across the top of the TV screen (and around the desert) reinforces the theme of colonization. Barbed wire is used to restrict the movements of colonized, enslaved, or jailed people. What are the odds of this being a coincidence?

Back to the dry terrain. The commercial strongly suggests that the big black ants (Africans/Black males) are carrying the Budweiser bottle (European king) across the dry terrain of Mother (Earth) Africa because Africans can't solve the problems of the thirsty Mother (Earth) Africa by themselves. They had to get a king from another country to solve their problems. The commercial suggests that the African needs the European to save him.

But as we know from history, the European king does not save Mother (Earth) Africa. Instead of bringing water to the dying land and people, the king pours alcohol into her, which only quickens the death of the people and the land. In many African cultures, libation, i.e., pouring alcohol onto the ground, is a sacred ritual that honors the ancestors. In this "Black Ant" commercial, however, the purpose of the alcohol libation is not to honor but to get drunk. Drunkenness makes it easier for the symbolic king to rape and steal the land and enslave the African people. This commercial is re-enacting scenes of rape, enslavement, and colonization straight out of history. And Brothers & Sisters, the cultural sabotage doesn't stop here.

On an urban level, the TV commercial carries a very profound message. The symbolism of Black men bringing a powerful white male into the community taps into the psycho-social behavior of the Black pimp. All bottles and cans of alcohol are phallic symbols. *The anthill is symbolic of female genitalia (i.e., vagina, womb).* The anthill is made out of earth, which is called Mother Earth, Mother Nature, the Cradle of Civilization, and the Great Mother. The commercial takes place in Mother Africa, which is called the Motherland. A mother, last time I checked, is a female. Furthermore, an anthill is, simply put, a hole in the ground. In street culture, females are often disrespectfully called *hoes*. When the big black ants (Black pimps) place the big bottle (white male) into the anthill (street "hoe"), we are looking at an act of sexual intercourse and prostitution. From an urban perspective, the commercial can be decoded as follows:

> ***Black males carry the white male into the community (Motherland) to intercourse with the Black "hoe" (anthill) in exchange for a drug, i.e., alcohol!***

On a basic level, nearly everyone who views the commercial knows that it's about sex. We further understand on a conscious level that the Euro king gives a "quickie," i.e., premature ejaculation, which implies that the king couldn't handle the power of the earth/Motherland/Black woman. That's why we laugh. What we don't know until we take the elements apart bit by bit is that this sex can either mean rape or prostitution. Is it still funny?

You might have to view the commercial a few times to see this, might even have to slow it down, but the big black ants jump into the anthill before the bottle slides in. So what? Sex sells, on conscious and subconscious levels. The symbolic African/Black males jump into the symbolic female vagina/womb *head first*. Don't laugh. The underlying theme is about gang rape. More than one ant/man jumps into the vagina/womb at one time. Alcohol is highly associated with rape, and rape is an act of violence. Needless to say, women suffer trauma and pain from the rape act. There's nothing sexy or funny about rape, yet it's being played out in prime time right before your eyes.

As the symbolic Black males return to the symbolic womb, the Cradle of Civilization, they become babies. After all, only a baby can live in a mother's womb.

Another way to look at it is this: Culturally speaking, whenever a Black male enters the womb (sexually) of his Black female, the lovers tend to use a unique speech pattern during the act of love making. She says, "Baby, oh baby." In response, the Black male calls the Black female "Mama." The womb of the African woman, or *wombman*, is so powerful, peaceful, and strong that it makes a full grown man as weak as a baby. Hello! It's common behavior for a man, when in the womb of the African wombman, to return to her breast repeatedly (perhaps for nutrition), as would a baby who feels love from its mama. Apparently, the Black male and Black female subconsciously understand that the safest place on the planet under the system of racism and white supremacy is inside the womb of a Black African wombman because

---

Hmmm . . .

The act of the symbolic Black male reentering the symbolic womb prompts this thought: "Jesus answered and said unto him, Verily, verily, I say unto thee, Except a man be born again he cannot see the Kingdom of God. Nicodemus saith unto him, How can a man be born when he is old? can he enter the second time into his mother's womb, and be born?"–John 3:3,4

outside, he's gonna catch hell.

After the Black males reenter the symbolic womb, the Budweiser bottle is forced into the womb of Mother Africa. It ejaculates, we see the bubbles. We know the bottle has been forced (this is rape) into the womb because, when you turn the sound completely down, you can clearly see that Mother (Earth) Africa is not dancing to the beat of K.C. and the Sunshine Band. *The jerking movements of the land suggests that she is bravely trying to throw the rapist off her body!*

Furthermore, the chemical that is released into her womb is alcohol, a drug that can kill the sexual response in a man. Alcohol is a toxic chemical. You can look at this at least one of two ways: (1) the African males become impotent (spiritually, physically, economically, intellectually, socially) at the release of the toxic chemical into the womb or (2) the African males, now babies that have returned to the womb, are killed via a chemical abortion. We are talking psychosocial abortion, physical abortion, or both. Furthermore, there are many big black ants (Black male babies) in the womb when the chemical is released. The hidden message here is about the genocide of African men and population control.

Remember when the bottle (phallus) goes into the anthill (womb)? Note that all the music in the video stops. You no longer hear the African drums and tribal chants. The message here is that in many African cultures, the communication has stopped. During the slavery era in America, Africans were forbidden to communicate through the drum or their native languages. When the drum was killed, so went a vital part of the African soul.

The sound that breaks the eerie silence is a cap being removed, i.e., the cap is popped. In street culture, to pop a cap means to ejaculate. ("Pop a cap" should not be confused

Hmmm . . .
The singers of "Do a little dance" and the symbolic king are both European. Could this be an accident? The use of K.C. and the Sunshine Band and that particular song reinforces our belief that the Black Ant commercial is about rape, not a harmless, funny sex act. Another interesting aside: there is a saying that goes, "you're not a man until you've had sex with a Black woman." During slavery, raping the African woman was a rite of passage for white males.

with "bust a cap," which means to shoot a person.) Immediately following the symbolic ejaculation, you hear the subliminal sound of a male sighing rather pleasurably. Now the elders in my family tell me that they often use small amounts of beer to help germinate (jumpstart) their soil. Could it be that the symbolic king is fertilizing, or impregnating, his rape victim?

The symbolic king has given his phallus a pet name, a common practice within white male culture (and possibly others). White males commonly refer to their penises as *dick, peter, pecker, cock, bone,* and *buddy*, just to name a few. Clearly, the king has chosen to call his phallus *Bud*. Now you know why the symbolic king's phallus (bottle) appears erect in the womb or genitalia of Mother (Earth) Africa. The slogan "This Bud's for you" appears across the bottom of the TV screen.

After the symbolic king ejaculates, we hear something that sounds like a zipper being unzipped. Then we hear "Do a little dance/make a little love/get down tonight." Clearly, these lyrics are the thoughts of the symbolic king. The symbolic king gives the big black ants (African/Black males) alcohol and tells them, through K.C., a cultural bandit group from the late '70s, to "Do a little dance," i.e., enjoy yourself, ignore what is happening to your African mother. He then tries to minimize his act of rape by telling his victim that they are "making a little love." Then he tells the rest of the village "get down tonight." What the symbolic king fails to tell them is that he will repeat this same traumatic behavior the next day.

The commercial also has an obvious Oedipal theme. (Oedipus is the guy who killed his father, then slept with his own mother.) All life comes from Mother Africa (if you didn't know, you'd better ask somebody). According to scientists, our most ancient mother was an African woman. Thus when the European king enters the womb of Mother (Earth) Africa, he too has returned to the womb of his original mother, the African woman or wombman. Because the commercial depicts him as intercoursing with his original mother (raping her), one can fairly classify him as a "mother f-r," just like Oedipus.

Recall that the terrain of Mother (Earth) Africa was depicted as dying of thirst. In a land where a severe drought is occurring, the people will try and get their water from underground water wells. A water well is simply a hole in the earth. The commercial strongly suggests that the anthill is actually a water well and the symbolic European king with the help of other Africans, poisoned the water supply by pouring the toxic chemical of alcohol into the well. This poisoning-AIDS, small pox, Eboly, etc.- of Mother (Earth) Africa strikes a cord of sorrow inside every African person on the planet.

When you say Budweiser, you've said it all, or, as Brother Neely Fuller said, "If you do not understand white supremacy (racism)-what it is, and how it works-everything else that you understand, will only confuse you."

Brothers & Sisters, Budweiser is culturally inappropriate to consume. This Bud's *not* for you, nor does Bud make you wiser.

An African-Centered Approach to 40oz Prevention
by Donna Marie Williams

"No fortress is impenetrable, just badly attacked."
–Jacqueline Butler, Associate Professor-Clinical Psychiatry, University of Cincinnati College of Medicine

Scripture says that the people perish for lack of knowledge. The *Art of War* advises that we know our enemy before we engage him in warfare. Thanks to Coach Powell, we're no longer ignorant of the underhanded ways that alcohol, specifically malt liquor, is being sold to our young people. We may not be wise to all the marketing schemes of the alcohol industry, we may not even know why these companies are specifically targeting Black youth (although I have my suspicions), but we do have enough information to start moving against the 40oz producers.

According to most of the research on prevention and youth development, the most effective strategies and programs are rooted in culture. The best way to fight the brewers on behalf of Black children is to implement African-centered programs that will stop the problems of alcohol experimentation and addiction before they have a chance to begin.

Years ago, somebody actually got paid to make up the slogan "just say no." As Judson Hixson, Senior Associate Director,

Midwest Regional Center for Drug-Free Schools and Communities, loudly preaches, rather than telling this generation to just say no to alcohol and other drugs, our best shot at helping them grow healthy, sane, and strong is by giving them something to say *yes* to.

Let's start with the young people who have the spark of life in them, the ones who want to grow up to create great music, art, scientific and technological inventions, the ones who want to help move the race a step higher. Let's go for the wannabes who might be straddling the fence of indecision and confusion. All they may need is a slight push on the right side to put them on the healthy path. Let's plant some fertile seeds within the so-called hard-core street culture. You never know who might be listening and who might be ready for the truth.

Coach and Shack have travelled the country for the past few years, talking about the 40oz scandal, revealing the hidden messages in the liquid dope that's being pushed like crack-cocaine into our communities, and without fail, the young people hear them. It never ceases to amaze me how, even with a primarily African-centered message, middle class kids in the suburbs, poor kids in the ghettoes, and farm kids in the rural Midwest really hear and resonate to the truth about the 40oz scandal. Coach and Shack have talked to Black kids, Caucasian kids, and Hispanic kids, male and female, and the response is always the same. They want to know more about how they've been conned, and they want to know how they can spread the message to their peers. Some even make the decision to stop sucking from the slavemasters' bottles of death, and that's the point of this message.

Created by Maulana Karenga in 1965, the Nguzo Saba provides a culturally relevant philosophical framework upon which our programs and strategies should be designed, developed, and implemented. Most of us have become familiar with the seven principles through Kwanzaa (celebration of the first fruits), the African American holiday that falls on December 26 through January 1. The seven principles of the Nguzo Saba are:

> **Umoja (unity)**-to strive for and maintain unity in the family, community, nation, and race.
> **Kujichagulia (self-determination)**-to define ourselves, create for ourselves, and speak for ourselves, instead of being defined, named, created by, and spoken for by others.
> **Ujima (collective work and responsibility)**-to build and maintain our community together and to make our sisters' and brothers' problems *our* problems and to solve them together.
> **Ujamaa (cooperative economics)**-to build and maintain our own stores, shops, and other businesses and to profit from them together.
> **Nia (purpose)**-to make as our collective vocation the building and developing of our community in order to restore our people to their traditional greatness.
> **Kuumba (creativity)**-to do always as much as we can, in the way we can in order to leave our community more beautiful and beneficial than when we inherited it.
> **Imani (faith)**-to believe with all our hearts in our people, our parents, our teachers, our leaders, and the righteousness and victory of our struggle.[100]

These seven principles are our battle cry. The alcohol industry is waging a vicious war for the minds and dollars of our young people. There have been many casualties. Only a strong, multifaceted counterattack will bring our young people back to sanity and health, and the alcohol industry to its knees.

Our fight must embrace who we are as a race. The beauty of the Nguzo Saba is that it updates ancient African ideals for a 21st century people. Our challenge to you is this: As you go to work in your communities, consider using the following philosophical framework to begin planning and implementation of programs. Check your existing programs against each and every one of the ideals. If things don't add up, something's wrong, and you'll need to reconsider how you've been doing business. Our programs will be different from community to community because Black people live in all kinds of situations: rural, urban, suburban, rich, poor, and indifferent. If we all adopt the principles of the Nguzo Saba, there will be unity among us, whether your program is based in the Mississippi Delta or in Harlem. It will be as if we're all walking on the same soil, breathing the same air, and feeling the warmth of the same sun. If I've learned nothing else in my work with Black youth I've learned this much: They will listen when you tell them the truth from their own unique cultural perspective. And is that so strange? Don't we all approach life through, as Zora Neal Hurston put it, our own cultural chemise?

The following adapts the seven principles to the cause of African American wellness and the total elimination of 40oz use among our young people.

NIA (PURPOSE)

We have one purpose, and that is to completely stop the spread of alcohol and other drug use among young people. We must spread the word from coast to coast. We must sing about it in our rap songs, gospel, and R&B tunes, and we must dance to it. *We must beat our drums to it.* A healthy culture will produce healthy people, and our children will grow to be sane, strong, and productive.

Umoja (unity) and
Ujima (collective work and responsibility)

As I see it, these two ideals go hand in hand. Our programs will fail if there is no unity among us. Queen Latifah rapped about it, the federal government pays lip service to it, but Brothers & Sisters, we've got to truly be about it. Collaboration has got to be more than just a buzzword. It's got to be our reality. With diminishing government and philanthropic resources, we're going to have to put aside turf wars, egos, and titles for the sake of our children. The alcohol industry may have their individual platoons called Stroh, Miller, and Anheuser-Busch, but together they form a powerfully funded army with a vast network of resources at their disposal. They may fight each other tooth and nail for market share, but you'd better believe, they are united on one primary point: to make you buy buy buy alcohol products. They would prefer for you to buy a competitor's product than to not buy any alcohol at all. But even with their money, political power, and distribution networks, the alcohol companies would be no match for a strong, united prevention front if only we'd wake up! Never forget that divided we fall, but working together, we have universal power. We have God on our side.

Kujichagulia (self-determination)

Who are our leaders? What is leadership? Which "leaders" have been doing our thinking for us? To make self-determination a reality in our communities, we must first redefine our antiquated concept of leadership. One of our most revered cultural values has been to submit to people in authority–in the pulpit, in the classroom, on TV, and in government. That's something the alcohol industry knows about us, and that's why the use of rap artists in malt liquor ad campaigns has been so devastatingly effective. In America's slavery/Jim Crow past, submission was a matter of survival, but today, following others whose personal agendas and political/corporate affiliations are largely unknown no longer serves us. A case in point are the zoning laws passed by our "leaders" that have allowed billboard companies to litter our streets and air space with alcohol and tobacco advertisements. The times demand that we all assume the mantle of leadership. It's not about ego tripping, but service. Kujichagulia leaders create their own visions, and they decide for themselves how to turn those visions into reality. Kujichagulia leaders have learned how to choose their friends and allies wisely, and their first consideration is always Black children. Kujichagulia leaders respect, appreciate, learn from, and work with other Kujichagulia leaders on behalf of Black children.

UJAMAA (COOPERATIVE ECONOMICS)

According to some estimates, our collective buying power ranks us the seventh largest nation in the world, so we've got the money. We don't need to accept dirty money from the alcohol industry any more. No more alcohol money for our charitable organizations, Black expos, and sporting and entertainment events. Look at what we can do when we put our minds, creativity, energy, and resources together with the goal of accomplishing a worthy vision. Black communities from coast to coast have magnificent churches that rival the great cathedrals, mosques, and temples of the world. We built those churches with our own hard-earned money. Now, look at how we waste our money on alcohol, cigarettes, and drugs. Surely our children rank higher than a bottle of OE. If we can build 10,000-seat churches, if we can keep Revlon and McDonald's in business, then we can pool our resources to (1) prevent the flow of dirty money into our communities and (2) create opportunities for our young people to heal and develop personally, socially, physically, economically, and academically.

Kuumba (creativity)

We are the people who made a delicacy out of hog guts. Surely we can eliminate alcohol and other drug use and addiction among our young people. We are the people who created jazz, the blues, gospel, and breakdancing. We invented the stoplight, heart surgery, and peanut butter. We have the unique godlike ability to create works of art and science, often out of nothing. If ever we needed the creativity of Black people, we need it now. The survival of an entire generation is at stake. Our agencies, organizations, institutions, and programs must not be so bureaucratic that they stifle the free flow of ideas. Our programs must be organic, living, breathing entities that respond quickly and effectively to youth and community needs. It's time for the rebels and geniuses to make their contributions to the cause of sobriety and social competence among Black youth!

Imani (faith)

We know how big the problems are, but we must refuse to allow feelings of futility and depression to overwhelm us. Wouldn't the alcohol industry just love to have us throw up the white flag of surrender and hand our children to them on a silver platter. I ain't giving up, and neither should you. The mere fact that you've hung with us until the end of this wild and wonderful book says that you're not a quitter. Our cause is righteous, and I know the Creator will open the windows of heaven to help us save Black children. The indigenous peoples of this land call their children "sacred ones," and so are ours. Our children carry on, not only our seed, but our hopes and dreams for the future.

They are designed to evolve us into greater beings, and they will stand a chance if we can keep them sober, sane, and healthy. I believe we can do it, and if all of us believe it and act like we believe it, all the slick con ads, all the dirty money laundering in the world won't be able to touch them.

Answers to the code hidden within the title of this book: '*N A* = N/A, or Not Applicable, i.e., alcohol is not applicable to the lives of Black people. '*N A* is also code for New African and Never Again.

ENDNOTES

The following dictionaries were used in *Message 'N A Bottle: The 40oz Scandal* (specialized dictionaries are given standard citations): *Webster's Third New International Dictionary* (Chicago: R.R. Donnelley & Sons Comapny, The Lakeside Press, 1961); *New Lexicon Webster's Dictionary*, 89th ed. (Danbury, CT: Lexicon Publications, Inc., 1992); *American Heritage Electronic Dictionary*, Houghton Mifflin Company, 1992 and *The American Heritage Dictionary*, 3rd ed., version 3.0A (WordStar International, 1993); *Reader's Digest Great Encyclopedic Dictionary* (Pleasantville, NY: The Reader's Digest Association, Inc., 1968).

Biblical scriptures cited are from the King James version, unless otherwise indicated.

Foreword

1. Bulhan, H.A., *Franz Fanon and the Psychology of Oppression* (New York: Plenum Press, 1985), p. 44.

2. Malcolm X, *On Afro-American History*, 3rd ed. (New York: Pathfinder, 1990), p. 37.

3. Adler, J., "The endless binge," *Newsweek*, Dec. 1994, p. 72.

Preface

4. Interview with Paul Kelly, Director of Alcohol and Drug Prevention, Bobby Wright Mental Health Center, and co-founder of the City-wide Coalition Against Tobacco and Alcohol Billboards (Chicago, Illinois), April 16, 1995.

5. *Forty Ounces & A Blunt*, Jan. 1994, inside cover.

6. Whitman, D., "The untold story of what really happened in the L.A. riot," *U.S. News & World Report*, May 31, 1993, p. 58.

ENDNOTES

Introduction

7. Drexler, M., "Tapping the youth market," *The Boston Globe Magazine*, Oct. 23, 1994, p. 8.

8. The Marin Institute for the Prevention of Alcohol and Other Drug Problems, Media Action Alert, July 23, 1991.

9. Griggs, F., "Teens turning to malt liquors," *Cincinnati Post*, April 19, 1993, p. 1A.

10. Ibid., p. 1A.

11. Jernigan, D., and Wright, P., *Making News, Changing Policy: Case Studies of Media Advocacy on Alcohol and Tobacco Issues* (Washington, DC: Center for Substance Abuse Prevention, no date), p. 115.

12. McMahon, E.T., and Taylor, P.A. *Citizen's Action Handbook on Alcohol and Tobacco Billboard Advertising* (Washington, DC: Center for Science in the Public Interest/Scenic America, January 1990).

13. Jung, C.G., *Man and His Symbols* (New York: Laurel, 1964), p. 3.

Chapter 1–Zima: Zomething Fishy

14. Bellant, R., *The Coors Connection: How Coors Family Philanthropy Undermines Democratic Pluralism* (Boston: South End Press, 1991), p. 67. (Quotation from *Rocky Mountain News*, Feb. 24. 1984, p. 64.)

15. Edmondson, A., "Zima's popular among teen-agers," *The Augusta Chronicle*, Feb. 28, 1995, p. 1.

16. Bellant, p. 1.

17. Ibid., p. 69.

18. Ibid., p. 66.

19. J.E. Cirlot, *A Dictionary of Symbols*, 2nd ed. (New York: Philosophical Library, Inc., 1974), p. 70.

20. *Handy Dictionary of the Bible,* ed. Tenney, M.C., 2nd ed. (Grand Rapids, MI: Zondervan, 1983), p. 39.

21. Promotional brochure for the United Methodist Church, Nashville, Tennessee.

22. Refer to these Eastern European dictionaries: *The Kosciuszko Foundation Dictionary: II Polish-English*, by Bulas, K., Thomas, L.L., and Whitfield, F.J. (New York: The Kosciuszko Foundation, NO DATE); *Serbocroatian-English Dictionary–Srpskohrvatsko-Engleski Recnik*, compiled by Benson, M. (University of Pennsylvania Press, 1971).

23. D.V. Perrott, *Concise Swahili and English Dictionary* (Chicago: NTC Publishing Group, 1965), p. 78.

24. Beverly Hall-Ogletree, "Many of Us and the Bottles Between Us." (Quoted with permission from Beverly Hall Ogletree.)

25. Phone conversation with staff member at Foote, Cone & Belding (San Francisco) advertising agency, Jan. 6, 1995. The staff member said that the offensive Zima commercial was revised.

26. Ketchum, K. and Mueller, L.A., M.D., *Eating Right to Live Sober* (New York: Signet, 1983), p. 68.

Chapter 2–Colt 45: The 40oz Harness

27. Freedman, A.M.,"Heileman tries a new name for strong malt," *Wall Street Journal*, May 11, 1992, p. B1.

28. Anthony Browder, *From the Browder File: Twenty-two Essays on the African American Experience* (Washington, DC: Institute of Karmic Guidance, 1989) p. 67.

ENDNOTES

29. *Advertising Age*, Nov. 9, 1984, as reported in the Malt Liquor Fact Sheet. Compiled and disseminated by The Marin Institute for Prevention of Alcohol and Other Drug Problems.

30. Storm, J., "Stand-up comedian Mark Curry, a prominent teacher on comedy," *Dayton Daily News TV Week*, July 11, 1993, cover story.

31. Lynch, William, "Let Us Make a Slave," otherwise known as "The Origin and Development of a Social Being Called the Negro." Reprinted by Alkebulanian Magazine.

32. Whitman, D., "The untold story of what really happened in the L.A. riot," *U.S. News & World Report*, May 31, 1993, p. 58.

33. Refer to McGee, V.J., *Through the Bible Commentary Series: The Prophecy Revelation Chapters 6-13* (Nashville: Thomas Nelson, Inc., 1991), p. 42.

34. Walker, B.G., *Women's Dictionary of Symbols and Sacred Objects* (San Francisco: Harper & Row, Publishers, 1988), p. 9.

35. Bullinger, E.W., *Number In Scripture* (Grand Rapids: Kregel Publications, 1967), p. 49.

36. Davidson, H.X, *Somebody's Trying to Kill You: The Psychodynamics of White Racism and Black Pathology*, vol. 1, rev. ed. (Kansas City: Ethos Growth and Development Publications, 1991), p. 5.

37. Liungman, C.G., *Dictionary of Symbols* (Denver: ABC-CLIO, Inc., 1991) p. 308.

38. Shabazz, I.A., *Symbolism, Holidays, Myths, and Signs*, 3rd ed. (Jersey City: New Mind Productions, 1991), p. 19.

39. Ketchum, pp. 90-91.

Chapter 3–Olde English 800: The British Are Coming!

40. DuBois, W.E.B., "Drunkeness," *The Crisis*, 1928, p. 35.

41. "Malt Liquor and Inner City Communities: Key Facts," The Marin Institute for Prevention of Alcohol and Other Drug Problems.

42. Williams, R., EdD., *They Stole It, But You Must Return It* (Rochester, NY: HEMA Publishing, 1991), p. 87.

43. Williams, E., *Capitalism and Slavery* (Chappel Hill, NC: University of North Carolina Press, 1994), p. 79.

44. Alderman, C.L., Rum, *Slaves and Molasses: The Story of the New England's Triangular Trade* (Great Britain: Bailey Brothers & Swinfen Ltd., 1974), p. 21.

45. Whitman, p. 35.

46. "Born again," *Source*, Feb. 1993, p. 41.

47. Asante, M., and Mattson, M., *Historical and Cultural Atlas of African Americans* (New York: Macmillan Publishing Company, 1991), p. 26.

48. Hayter, T., *The Creation of World Poverty* (Boulder, CO: Westview Publishers, 1981), p. 20.

49. *Collier's Encyclopedia*, ed. Bahr, L.S. (New York: P.F. Collier & Sons, Ltd., 1993) pp.587-593.

50. Davidson, B., *The African Slave Trade* (Boston: Little, Brown and Company, 1980), p. 81.

51. DuBois, W.E.B., *The Suppression of the African Slave Trade to the United States of America–1638-1870* (New York: Schocken Books, 1969), p. 3.

52. Cirlot, pp. 67-68.

ENDNOTES

53. Hall-Ogletree, B., *Making the Right Choices: A Drug Prevention Curriculum for African American Children and Adolescents* (Dayton: B Educational Services, 1990), p. 7.

54. *Dictionary of American Slang*, ed. Chatman, R.L. (New York: Harper & Row, 1987), p. 91.

55. Bullinger, p. 200.

56. Jones, D., *Culture Bandits II* (Philadelphia: Hikeka Press, Inc., 1993), p. 199.

57. *Dictionary of American Slang*, p. 451.

58. Baus, H.M., *The Master Crossword Puzzle Dictionary*, 1st ed. (Garden City, NY: Doubleday & Company, Inc., 1981), p. 1535.

59. *The American Heritage College Dictionary*, p. 1416.

60. Cirlot, p. 343.

61. *Webster's Third New International Dictionary*, p. 2392.

62. *New Lexicon Webster's Dictionary*, p.1033.

Chapter 4–Red Dog: That Low Down Dirty Dog

63. Roeper, R., "Label upholds beer's 'bad dog' reputation," *Chicago Sun-Times*, March 23, 1995, p. 11.

64. Teinowitz, I., and Ingrassia, J.,"Miller barks up new tree with Red Dog," *Advertising Age*, Sept. 19, 1994.

Chapter 5–ST Ides: Satan 'N A Bottle

65. Marinucci, C., "Malt liquors–rapper ads changing tone," *San Francisco Examiner*, Dec. 15, 1991, p. E-1.

66. Greer, J., "Ad agency tackles new role in rollout of malt liquor," *San Francisco Chronicle*, June 12, 1987, p. 47, column 1.

67. Marinucci, p. E-1.

68. Liungman, p.306.

69. Greer, p. 47, column 1.

70. Shabazz, p. 19.

71. *Handy Dictionary of the Bible,* p. 44.

72. Compiled by Sutton, W.J., *The New Age Movement; and The Illuminati 666* (Institute of Religious Knowledge, 1983), p. 23.

73. Biedermann, H., *Dictionary of Symbolism: Cultural Icons and the Meanings Behind Them,* trans. by Hulbert, J. (New York: Penguin Group, Penguin Books, USA, Inc., 1994), p. 207.

74. Liungman, p. 191.

75. Biedermann, p. 252.

76. Becker, U., *The Continuum Encyclopedia of Symbols,* trans. by Garmer, L.W. (New York: Continuum, 1994), p. 225.

Chapter 6–Big Jug: The False Prophet

77. Akbar, N., lecture to school teachers in Pittsburgh, PA., Sept. 1991.

78. Key, W.B., *The Age of Manipulation: The Con in Confidence, The Sin in Sincere* (Lanham, MD: Madison Books, 1989), p. 121.

79. Kotulak, R., "How brain's chemistry unleashes violence," *Chicago Tribune,* Dec. 13, 1993, pp. 1, 8.

80. *Biological Rhythms, Mood Disorders, Light therapy, and*

the Pineal Gland, ed. Shafii, M., M.D., and Shafii, S.L., R.N., B.S.N. (Washington, DC: American Psychiatric Press, Inc.), p. 48. For an African-centered perspective on seratonin, melanin, and the connection between the two, read Richard D. King, M.D., *African Origin of Biological Psychiatry* (Germantown, TN: Seymour-Smith, Inc., 1990); Frances Cress Welsing, M.D., *The Isis (Yssis) Papers: The Keys to the Colors* (Chicago: Third World Press, 1991); Carol Barnes, *Melanin: The Chemical Key to Black Greatness* (Houston: C.B. Publishers, 1988); and Jeral Constance Muhammad, *The "Golden Key": Melanin, Unveiling the Missed-Story to the Black Masses* (Des Moines, Larej Enterprises, 1990).

Chapter 7–Schlitz: The Bull God Osiris Resurrected?

81. Cirlot, p. 33.

82. Welsing, p. 70.

83. Michael Porter, Vice President, Stroh Brewery. Interviewed by Stossel, J., "Buying the hard sell," ABC's *20/20*, Jan. 15, 1993.

84. Welsing, p. 70.

85. Guard, S., "For the record," *Sports Illustrated*, Sept. 29, 1992, p. 69.

86. Taylor, P., "Callous disregard," *Sports Illustrated*, April 6, 1992, p. 9.

87. Winstone, H.Z.F., *Uncovering the Ancient World* (New York: Facts on File Publications, Inc., 1986) p. 221.

88. As reported in *Alcohol and Other Drug-related Violence Prevention*, The Center for Substance Aubse Prevention, 1989.

89. Butterfield, J., "Sex: Girls are worried," *Chicago Sun-Times USA Weekend*, August 12-14, 1994, p. 10.

90. Winstone, p. 221.

91. See Jackson, J.G., *Pagan Origins of the Christ Myth* (Austin, TX: The American Atheist Press, NO DATE); ben-Jochannan, A., *The African Called Rameses (The Great) II and the African Origin of "Western Civilization"* (Harlem: The Third Eye, Inc., 1990); Massey, G., *The Historical Jesus and the Mythical Christ* (Brooklyn: A&B Publishers, NO DATE); and Anyike, J.C., *Historical Christianity African Centered* (Chicago: Popular Truth, Inc., 1994).

92. Anyike, p. 109.

93. Massey, p. 75.

94. Jackson, J.G., *Man, God, and Civilization* (Secaucus, NJ: Citadel Press, 1972), p. 129.

Chapter 8–Budweiser: The Final Confrontation

95. Fuller, Jr., N., *The United Independent Compensatory Code/System/Concept*, 1971, p. 1.

96. Jung, p. 3.

97. Lecture given by Frances Cress Welsing, M.D., in Dallas, Texas.

98. *What's In A Name? Surnames of America* (New York: Harcourt Brace Jovanovich, Inc., NO DATE), p. 887-88, 428.

99. *Advertising Age*, May 29, 1995, p. 3.

An African-Centered Approach to 40oz Prevention

100. Madhubuti, H.R., *Kwanzaa: A Progressive and Uplifting African-American Holiday* (Chicago: Third World Press, 1987), pp. 5-6.

SUGGESTED READINGS*

I. Classical Sources

Ani, M. Yurugu: *An African-Centered Critique of European Cultural Thought and Behavior.* Trenton, NJ: Africa World Press, Inc., 1994.

Anyike, J.C. *Historical Christianity African Centered.* Chicago: Popular Truth, Inc., 1994.

Ben-Jochannan, Y.A.A. *African Mothers' Western Civilization.* New York: Alkebu-lan Books, 1971.

Ben-Jochannan, Y.A.A. *Black Man of the Nile and His Family.* New York: Alkebu-lan Books, 1981.

Cruse, H. *The Crisis of the Negro Intellectual: A Historical Analysis of the Failure of Black Leadership.* New York: Quill, 1967.

Diop, C.A. *African Origins of Civilization.* New York: Lawrence Hill and Company, 1974.

Diop, C.A. *Cultural Unity of Black Africa.* New York: Lawrence Hill and Company, 1987.

Jackson, J.G. *Introduction to African Civilization.* New York: University Press, 1970.

James, G.G.M. *Stolen Legacy.* San Francisco: Julian Richardson Associates, 1976.

Selected readings from the Classical Sources and Foundation Readings sections were primarily compiled for *An African Centered Model of Prevention for African-American Youth at High Risk.* ed. Goddard, L.L. Center for Substance Abuse Prevention, U.S. Department of Health and Human Services, Washington, DC.

Rogers, J.A. *Africa's Gift to America.* New York: Helja M. Rogers Publisher, 1961.

Rogers, J.A. *The World's Greatest Men of Color.* Vols. I and II. New York: MacMillan, 1973.

Snowden, Jr., F. *Blacks in Antiquity.* Cambridge, Mass.: Harvard University Press, 1970.

Williams, C. *The Destruction of Black Civilization, Great Issues of a Race From 4500 B.C. to 2000 A.D.* Chicago: Third World Press, 1974.

II. Foundation Readings

Asante M.K. *Afrocentricity: The Theory of Social Change.* Buffalo, NY: Amulefi Publishing Co., 1980.

Asante, M.K. *The Afrocentric Idea.* Philadelphia: Temple University Press, 1987.

Asante, M. Kemet, *Afrocentricity and Knowledge.* Trenton, NJ: African World Press, Inc., 1990.

Asante, M., and Asante, K. *African Culture: the Rhythms of Unity.* Trenton, NJ: African World Press, Inc., 1990.

Drake, St. C. *Black Folk Here and There: an Essay in History and Anthropology.* Vol I. Los Angeles: Center for Afro-American Studies, University of California, 1987.

Drake, St. C. *Black Folk Here and There: An Essay in History and Anthropology.* Vol II. Los Angeles: Center for Afro-American Studies, University of California, 1990.

DuBois, W.E.B. *Black Folks: Then and Now; an Essay in the History and Sociology of the Negro Race.* New York, NY: Octagon Books 1970 (1935).

SUGGESTED READINGS

Erny, P. *Childhood and Cosmos: The Social Psychology of the black African Child.* New York: Independent Publishers Group, 1973.

Karenga, M. *The Husia.* Los Angeles: University of Sankore Press, 1986.

Karenga, M. *Introduction to Black Studies.* Los Angeles: University of Sankore Press, 1986.

Karenga, M., ed. *Reconstructing Kemetic Culture: Papers, Perspectives, Projects.* Los Angeles: The University of Sankore Press, 1990.

Karenga, M., and Carruthers, J.H., eds. *Kemet and the African World View, Research, Rescue and Restoration.* Los Angeles: University of Sankore Press, 1986.

King, K.; Dixon, V.; and Nobles, W.W., eds. *African Philosophy: Assumptions and Paradigms for Research on Black Persons.* Los Angeles: Fanon Center Publication, Charles R. Drew Postgraduate Medical School, 1976.

Nobles, W.W. *African Psychology.* Oakland, Calif.: Black Family Institute, 1986.

Richards, D. *Let the Circle Be Unbroken.* New York: DA Publishers, 1989.

Van Sertima, I., ed. *Journal of African Civilizations.* Vol. 4, No. 1, New Brunswick, NY: Transaction Books, 1982.

Van Sertima, I., ed. *African Presence in Early Europe.* New Brunswick, NJ: Transaction Books, 1985.

Van Sertima, I., ed. *Egypt Revisited.* New Brunswick, NJ: Transaction Books, 1985.

Van Sertima, I. *They Came Before Columbus.* New Brunswick, NJ: Transaction Books, 1985.

Warfield-Coppock, N. *Afrocentric Theory and Applications. Volume 1: Adolescent Rites of Passage.* Washington, DC: Baobob Associates, 1990.

Welsing, F.C. *The Isis Papers: They Keys to the Colors.* Chicago: Third World Press, 1992.

III. Readings in Media, Advertising, and Related Topics

"The News Media and the Disorders," *The Kerner Report: The 1968 Report of the National Advisory Commission on Civil Disorders.* New York: Pantheon Books, 1968.

Bagdikian, B.H. *The Media Monopoly.* 3rd ed. Boston: Beacon Press, 1990.

Chomsky, N. *Necessary Illusions: Thought Control in Democratic Societies.* Boston, MA: South End Press, 1989.

DeMoss, R.G. *Learn to Discern.* Grand Rapids, MI: Zondervan Publishing House, 1992.

Jung, C.G. *Man and His Symbols.* New York: Laurel, 1964.

Key, W.B. *Subliminal Seduction: Ad Media's Manipulation of a Not So Innocent America.* New York: Signet Books, 1973.

Mander, J. *Four Arguments for the Elimination of Television.* New York: Quill, 1978

McLuhan, M. *Understanding Media: The Extensions of Man.* New York: Signet Books, 1964.

McMahon, E.T., and Taylor, P.A. *Citizen's Action Handbook on Alcohol and Tobacco Billboard Advertising.* Washington, DC: Center for Science in the Public Interest and Scenic America, 1990.

Parenti, M. *Inventing Reality: The Politics of the Mass Media.* New York: St. Martin's Press, 1986.

SUGGESTED READINGS

Schiller, H.I. *The Mind Managers.* Boston: Beacon Press, 1973.

Winn, M. *Unplugging the Plug-in Drug.* New York: Penguin Books, 1987.

Coach Alfred Powell is available to speak to young people and adults on the 40oz scandal. He and his partner, Marlon "Shack" Shackleford, also conduct workshops on a wide range of topics, including youth empowerment, building self-worth, and male-female relationships. For more information on materials and workshops, contact:

Renaissance Communications

1507 E. 53rd St., #257

Chicago, IL 60615

312/928-EDIT (3348)